Hybridization, Diagnostic and Prognostic of Proton Exchange Membrane Fuel Cells

Series Editor
Alain Dollet

Hybridization, Diagnostic and Prognostic of Proton Exchange Membrane Fuel Cells

Durability and Reliability

Samir Jemeï

iSTE

WILEY

First published 2018 in Great Britain and the United States by ISTE Ltd and John Wiley & Sons, Inc.

ISTE Ltd
27-37 St George's Road
London SW19 4EU
UK

www.iste.co.uk

John Wiley & Sons, Inc.
111 River Street
Hoboken, NJ 07030
USA

www.wiley.com

Library of Congress Control Number: 2018954267

British Library Cataloguing-in-Publication Data
A CIP record for this book is available from the British Library
ISBN 978-1-78630-167-3

Contents

Foreword

When reflecting on what our energy future should be, specialists in the field use various terms: abundance, sustainability, renewable within human life span, environmental friendliness, accessibility to all in every corner of the planet and multiplicity of application contexts. Hydrogen meets all these requirements. It is indeed the most abundant element in the universe (75% in mass and 92% in number of atoms). It would then seem logical and even obvious to consider it an interesting component in our future energy mix. Unfortunately, since it is rarely found on Earth in its dihydrogen form, it needs to be produced. The hydrogen thus obtained can be characterized as a "hydrogen energy vector", which forms a natural duality with the electricity vector (passages from one vector to another are obtained either by water electrolysis or by using a hydrogen fuel cell). Because of this duality, its use is rendered particularly easy, while introducing an intrinsic capacity to limit the environmental impact of the full energy chain (indeed, hydrogen can and should be produced without fossil fuels). The commonly used term is hydrogen-energy, in which the suffix clearly highlights the goal to be obtained.

In this rapidly evolving energy context, disruptive technologies (among which is hydrogen-energy) are worth being studied in depth. In this book, we rightly propose a systemic view of technological, economic and social stakes related to this energy vector. Owing to his unique experience of approximately 20 years in this field, Dr Samir Jemei goes beyond this scope. He draws on this experience to identify the major scientific challenges to be met in order to bring hydrogen fuel cells to the level of technological maturity, which would enable them to become fully competitive commercial products.

Therefore, this work offers interested readers an in-depth view on these subjects, focusing on two major issues: hybridization of hydrogen power generators and identification of their (current and future) state of health. The results of numerous scientific studies conducted in recent years are thus summarized here. The advantages presented by using advanced signal processing methods for real-time control of energy flows in hydrogen fuel cells is particularly worth noting, together with artificial intelligence-based approaches for the state of health diagnosis and remaining lifetime prognosis for these electrochemical generators.

Despite the complexity inherent in the studied systems, this book provides an astonishingly didactic value for the layperson. The reader can thus become progressively acquainted with the stakes, methods and adopted strategies (What are the study requirements? What are the specific constraints associated with the hydrogen fuel cell? What are the various possible approaches? What are the principles? What measurements enable a choice? What are the expected results?). The very rich bibliography provides readers with the opportunity to expand their exploration of a specific field of study. A further teaching benefit stems from the approaches used, which are systematically evaluated, thanks to experimental data (generated by experiments conducted on hydrogen fuel cell testing bench or on full experimental systems (particularly hydrogen-powered electric vehicles)).

With a skillful combination of scientific theory and industrial practice, this book is an essential reference for all the engineers and researchers working in the field of integration of hydrogen fuel cell systems, either in environmentally friendly stationary applications or in clean transport.

Professor Daniel HISSEL
University of Burgundy Franche-Comté
FEMTO-ST Institute (CNRS)
FCLAB Federation (CNRS)
Belfort
May 2018

Introduction

I.1. Subject matter

Global emissions of greenhouse gases covered by the Kyoto protocol reached approximately 49 billion tonnes of CO_2 equivalent in 2010. They increased by 80% between 1970 and 2010 and by 30% since 1990 [FRA 15]. The large majority of these emissions are obviously related to the combustion of fossil energies. The resulting pollution and the irreversible depletion of fossil resources should stimulate a reflection on our global energy system, which seems on its last legs. Many measures that have been taken through various programs, such as sustainable development, Horizon 2020 and even COP21, aim to decrease greenhouse gas emissions, reduce energy consumption, stabilize global warming, diversify primary energy sources or even develop renewable energies. Taking into account these elements forces the system into a new energy transition and tends to increase the share of renewable energies in the energy mix.

Power generation using wind and solar energies is reaching maturity, with ever-increasing production. Nevertheless, irregularity in the availability of these renewable energies gives rise to many problems in terms of power grid management. Managing a grid that day by day involves slightly more intermittent energies requires availability of reserve storing and release in order to cope with the differences between supply and demand in a given territory. It is indeed important to be able to deal with various time scales: daily (day/night), weekly (weekend/weekdays) and seasonal (summer/winter). Yet, it proves that hydrogen is ideal for storing this renewable energy. It can be produced using water electrolysis, which is itself fed by renewable energies. The resulting gas is then stored under various forms in order to be used at the appropriate time. Hydrogen can be combined with natural gas in existing networks, can be directly used in the industry or can even be used to again generate electric power by means of a fuel cell.

Since the fuel cell (FC) is at the center of the research activities presented in this book, it is worth our full attention. Indeed, this electrochemical generator converts the chemical energy of a reaction between a fuel (hydrogen) and an oxidizer (oxygen in the air) to generate electricity, heat and water. It can be used for stationary or nomadic applications or in transport. Moreover, in order to decarbonize the transport, the use of the hydrogen energy vector in combination with a fuel cell becomes an increasingly mature solution. As a proof, many global car manufacturers are manufacturing and marketing their own fuel cell vehicles. The most recent example is that of the Japanese manufacturer Toyota, which sold approximately 1,000 units in 6 months and is expecting to reach 30,000 fuel cell vehicles per year from 2020.

In Japan, except for stationary applications, more than 200,000 fuel cell systems have so far been installed in private homes to generate electricity and heat.

Hydrogen can be used in all sectors to store intermittent energy, supply mobile applications, meet the needs of an isolated site, which is disconnected from the power grid, propel a car, railway, naval or space flight transportation means and so on. In 1874, in his book *The Mysterious Island*, Jules Verne had predicted without an artificial intelligence algorithm that: *"water will one day be employed as fuel, that hydrogen and oxygen which constitute it, used singly or together, will furnish an inexhaustible source of heat and light, of an intensity of which coal is not capable"* [VER 74]. Half a century later, the rise of a hydrogen society is conceivable.

This book essentially deals with the problems related to fuel cell technology for energy generation. Even though the FC has experienced much progress in the last 20 years, the technical challenges have not yet been fully solved. The research work described in this book falls within this context, drawing on the works that relate to electrical engineering and, to a lesser extent, to the automation field.

I.2. Chapter breakdown

The works presented are centered on three axes organized in four chapters.

The first chapter starts with a brief overview of the current energy model. A review of the solutions that could enable the decarbonization of the energy mix is then provided. Among these solutions, a proposed focus is on the hydrogen vector, which is closely linked with the fuel cell generator. This chapter ends with the presentation of the fuel cell and its applications. Potential applications for transport and for the stationary field are highlighted. Finally, advantages and drawbacks of this technology are described. The FC presents problems related to cost, efficiency, lifetime, integration and hydrogen storage, but it has the advantage of releasing few or no pollutants, being silent, having high energy efficiency and offering substantial lifetime and reduced cost. Nevertheless, the FC is a multi-physical generator that becomes difficult to understand when considered with its auxiliaries.

The second chapter presents various existing types of FCs and identifies two technologies, namely the PEMFC (proton exchange membrane fuel cell) and the SOFC (solid oxide fuel cell), which are considered to be the most promising solutions for transportation and stationary applications. Regardless of the technology used, many auxiliaries are needed to operate the FC and it is important to control the whole system. Hence, all the auxiliaries are described and particular attention is given to the compressor unit that allows the air supply of the PEMFC, which is one of the largest energy consumers of the system (about 10–15%). With regard to the SOFC, which mainly differs from the PEMFC by its operating temperature (above 800°C), an integrated power generator is presented. For both technologies, experimental implementation is a key point, as it facilitates the understanding of how the FC and its system operate. These works can be considered the first axis, on which the research activities needed to obtain an augmented durability FC system rely. The second axis deals with the energy optimization of various hybrid energy sources, which enables an increase in FC lifetime. Finally, the third axis is dedicated to FC diagnosis and prognosis, which moreover enable the expansion of lifetime and an increase in generator reliability.

Thus, the third chapter deals with the hybridization of energy sources. If a transport application is considered, it is important to hybridize the FC with other sources, as the overall dynamics of the FC system is rather slow and causes difficulties in the case of rapid transients within the mission. Furthermore, in order to reduce the overall energy consumption of a vehicle, braking energy should be recovered, which is impossible with a single FC, as it is a non-reversible source. Finally, hybridization of this FC with energy-storing devices should be considered, in order to obtain high overall efficiencies. In such hybrid systems, an energy management strategy must be developed so that power demands are allocated to the right sources, at the right times, depending on the mission profile to be followed. Two approaches are proposed in this chapter. The first approach, which relies essentially on the wavelet transform combined with neural networks, enables the development of online energy management applied to a heavy hybrid vehicle composed essentially of a FC system, accumulators and ultracapacitors. It is worth noting that this energy distribution takes into account the frequency ranges of various sources using only the current and previous data of a univariate signal of the vehicle power demand. This will eventually allow real-time implementation as well as an increase in FC lifetime. The second approach relies on type-2 fuzzy logic and genetic algorithms. This solution is applied to a hybrid locomotive, composed of accumulators, ultracapacitors and a diesel engine, in order to minimize fuel consumption. The results show that the characteristics and dynamics of various sources have been taken into account thanks to artificial intelligence tools. High-performance energy management tools are proposed here, allowing optimal use of sources. Nevertheless, our efforts should focus on FC lifetime and reliability, in order to have a hybrid system with even higher performances.

For this purpose, FC diagnosis and prognosis have been developed in the fourth chapter. These two disciplines should enable us to reach our objective of increasing the lifetime and reliability of the FC. Diagnosis methods are used to identify the origin of failure and to determine the FC state of health, thus allowing for decision-making that ensures proper operation of the generator. PHM (Prognostics & Health Management) helps to predict the evolution of FC behavior in order to estimate a future failure. There are many approaches to the development of diagnosis and prognostic tools. Nevertheless, in order to find effective solutions, first it is important to have in-depth knowledge of the degradation mechanisms of the FC and its system. Thus, in the first part of Chapter 4, degradation mechanisms and the failures that the FC and its system may present are reviewed. In the second part, two diagnosis methods are presented. The first one (data-based) is a supervised classification method called "k-nearest neighbors". The second (signal-based) relies on a wavelet transform approach. These approaches have in particular enabled the diagnosis of various FC system failures that may occur with good classification rates reaching 90%. As will be shown, it is also possible to estimate the FC state of health using the wavelet transform coupled with energy indicators. Finally, in the last part of this chapter, methodologies related to the data-based prognostic are developed. The tools used to monitor the state of health and estimate the future behavior of the FC are based on neural networks. The first works are carried out using the ANFIS (Adaptive Neuro-Fuzzy Inference System). This model enables us to predict the behavior of voltage evolution. Nevertheless, the learning bases prove to be substantial. To reduce the required data and achieve better prediction performances, we have subsequently used the ESNs (echo state networks), which are new systems of neural networks. The conventional neural networks are often costly in terms of computation time due to algorithmic complexity. For the ESNs, algorithmic complexity is replaced by structural complexity, so that the learning phase is faster than that of conventional neural networks. Thanks to the use of ESNs of genetic algorithms and the wavelet transform, the results obtained are very encouraging. Indeed, they allow us to predict cell voltages until the end of life of the FC with an error below 10%.

1

Fuel Cells: the Path Towards Hydrogen Revolution

1.1. Introduction

In the 21st Century, it has become nearly impossible to do anything without means of transportation, communication or energy generation, which are expected to be increasingly efficient. These three concepts are present in many fields of activity of everyday life, such as telecommunications, information technology (thermal and/or electric) and energy supply for domestic or industrial use or for transportation (private, railway, maritime or air). These various technological means must have very high efficiency and release a minimum amount of pollutants, in order to increase the availability of our fossil energy resources and preserve our planet. To date, more than 80% of the primary energy supply relies on fossil fuels [BAD 13], which leads us, on the one hand, to a supply breakdown in the very near future and, on the other hand, to an environmental catastrophe. Indeed, the most significant and immediate risk relates to the climate change caused by the emissions of greenhouse gases due to human activities. The use of fossil fuels generates CO_2, which is the main man-made greenhouse gas contributing to global warming. In order to avoid an increase in Earth's average surface temperature by more than 2°C, human CO_2 emissions should be halved by 2050 globally [INT 15]. For this purpose, the European Union has set various objectives in terms of climate change and energy until 2020. It is the "three times twenty" rule that should enable the start of genuine energy transition:

– decrease in greenhouse gas emissions by 20% (even by 30%, if possible) compared to 1990;

– increase in energy efficiency by 20%;

– up to 20% use of energy from renewable sources for primary energy supply.

Firstly, this chapter will provide a brief overview of global energy, going through various possibilities for the future.

Secondly, the focus will be on the hydrogen vector and on the need to store it.

Finally, the fuel cell generator and its applications will be presented.

1.2. Energy: a global view

Global energy consumption has rapidly increased in the recent past, relying essentially on fossil energies. This explains why significant progress has been made over the past few years by developing cleaner and more efficient production technologies. Nevertheless, agreed efforts are not sufficient to avoid a 2°C increase in the Earth's average surface temperature. This requires an "energy mix" that will enable the preservation of resources and the environment, while at the same time meeting the demand.

1.2.1. *An obsolete energy model*

Energy demand is steadily increasing due to the requirement to meet the energy needs of increasing population. As previously mentioned, fossil energy reserves will be depleted and alternative energy sources need to be found. It should be recalled that in 2010, 80% of the global primary energy generation relied on fossil energies, 10% on hydroelectric energy, 6% on nuclear energy, 2% on biomass energy and only 1% on other renewable energies [INT 12]. Unless measures are taken, the carbon dioxide content in the atmosphere could increase by 30% in 20 years, while focusing on the development of other energy resources that generate less carbon dioxide waste, which could help us reverse the trend and decrease the emissions by 30% by 2035 (Figure 1.1). Examples of resources or technologies that release the lowest rates of CO_2 include solar, wind, hydraulic and nuclear energy (Figure 1.2). It is thus essential to propose a combination of renewable energies that can satisfy our various preoccupations and, in particular, to further develop decarbonized energies.

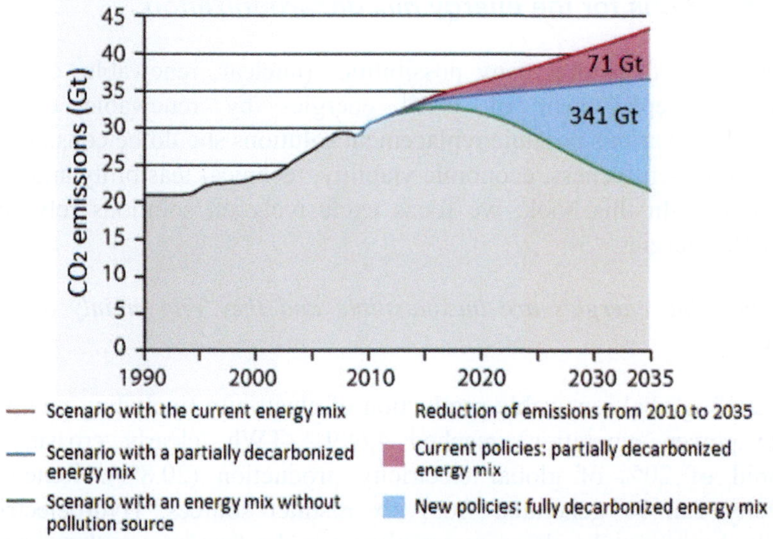

Figure 1.1. *Scenarios of the evolution of CO_2 emissions [INT 11]. For a color version of this figure, see www.iste.co.uk/jemei/cells.zip*

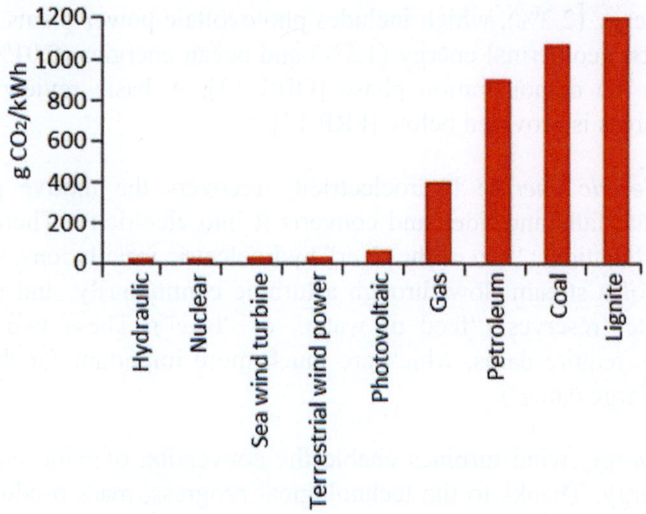

Figure 1.2. *Direct emissions + analysis of CO_2 life cycle by energy source [BER 14]*

1.2.2. *Solutions for the energy mix decarbonization*

Nowadays, there are many possibilities (nuclear, renewable, etc.) that enable the replacement of fossil energies by renewable energies. Nevertheless, various possible replacement solutions should be considered in terms of competitiveness, economic viability, technical feasibility and social acceptability. In this book, we focus exclusively on solutions relying on renewable energies.

"*Renewable energies are inexhaustible and they rely mainly on solar energy.*"

In 2012, global renewable production of electricity (including pump and turbine power generation) reached 4,699.2 TWh, clearly crossing the threshold of 20% of global electricity production (20.8%). Renewable electricity can be generated from six distinct sources. Hydroelectricity (including pump and turbine power plants) is the first among them, with a 78% contribution in 2012. Wind energy, which from 2009 has become the second common source of renewable energy, and now represents 11.4% of all renewable energy. It outperforms biomass, which covers solid biomass, liquid biomass, biogas and renewable domestic waste (6.9%). It is followed by solar energy (2.2%), which includes photovoltaic power plants and solar steam plants, geothermal energy (1.5%) and ocean energies (0.01%), which are still in the demonstration phase [OBS 13]. A basic review of these various sources is provided below [FRE 13].

Hydroelectric energy: hydroelectricity recovers the motive power of streams, waterfalls and tides and converts it into electricity. There are two types of installations: "run of the river" hydroelectric installations, which run all or part of a stream flow through a turbine continuously, and those that require water reserves ("feed of water" or "lake"). These two types of installations require dams, which are much more important for the second category ("large dams").

Wind energy: wind turbines enable the conversion of wind energy into electric energy. Thanks to the technological progress, mass production and accumulated experience, wind energy should become the largest contributor to electricity generation in the decades to come.

Biomass: there is a considerable difference between biomass and other renewable energy sources, to the extent that it takes the form of a fuel that can be stored and used to generate electricity as needed, similarly to fossil fuels. Nevertheless, biomass is often limited by the energy density of the stored fuel. The following are the three basic technologies for converting biomass energy into electricity, the primary fuel being solid-state biomass:

– direct combustion;

– gasification;

– pyrolysis.

Solar energy: solar power plants convert solar radiation into electricity or heat, depending on the technology employed.

Photovoltaic solar energy uses photovoltaic modules to generate electricity, which is then fed into power grids.

Solar thermal energy generates heat, which can be used for domestic heating or hot water generation.

Finally, thermodynamic solar energy generates electricity by heat production.

Geothermal energy: geothermal energy or "heat of the Earth" covers all the applications, enabling the recovery of the heat contained in the subsoil or in the water table (the higher the temperature of the ground and of the underground water, the closer one gets to the center of the Earth). Depending on the application, the calories thus recovered serve to generate heat and/or cold or electricity [NOT 17].

Ocean energies: ocean energy conversion includes all the technologies that enable energy generation, particularly electricity, based on the various resources of the ocean environment, such as swell, currents, tides, temperature gradient and wind (sea wind turbines) [NOT 17].

The European objective, which sets the share of renewable energies at 20% (23% for France) in the final gross consumption of energy until 2020, involves a strong development of intermittent sources such as wind or solar energies, whose generation profiles are sometimes out of phase with the profiles of electric energy consumption [GRH 16].

It is therefore essential to develop flexibility vectors that optimize the connection between energy generation and consumption. The hydrogen vector is one of the relevant candidates to be used as a flexibility solution.

1.3. Hydrogen vector

Hydrogen does not exist on Earth in its natural state. This element is the lightest and one of the most abundant in nature. It is necessarily linked to other elements such as dioxygen or carbon. Hydrogen can be generated by extracting and isolating the molecule. Water electrolysis is another solution. The dissociation of water molecule requires the same amount of energy that is recovered when hydrogen and oxygen are recombined. Hence, hydrogen is not a resource, but an energy vector.

1.3.1. *Hydrogen production*

Hydrogen can essentially be produced from three large sources, namely fossil, nuclear and renewable energies. Nowadays, half of the global hydrogen production results from methane reforming, which is the most economic source, and the other half is obtained from coal or petroleum. Electrolysis, which is an expensive technique, accounts for only 4% of the global hydrogen production, particularly when very pure hydrogen is needed. The supply of refineries and the fabrication of ammonia constitute 85% of the outputs [BAD 13].

1.3.1.1. *From fossil energies*

Hydrogen production in industrial applications relies on methane (CH_4), which is the main component of natural gas. The method used for this purpose is reforming. It is a chemical reaction that breaks the hydrocarbon molecules under the action of heat and releases hydrogen (dihydrogen, more precisely). There are two types of reforming:

– *Steam reforming* of natural gas is the most common process. Mixing methane and water vapor at high temperature (about 850°C) triggers the endothermic reaction (equation [1.1]):

$$CH_4 + H_2O \rightarrow CO + 3H_2 - \Delta H^0 \qquad [1.1]$$

with $\Delta H^0 = 252.3\ kJ.mol^{-1}$

A secondary reaction [1.2] called water–gas "shift reaction" (WGS reaction) enables the elimination of CO:

$$CO + H_2O \rightarrow CO_2 + H_2 - \Delta H^0 \qquad [1.2]$$

– *Reforming by CPOx* of the natural gas or catalytic reforming by partial oxidation (CPOx: catalytic partial oxidation) involves reaction [1.3].

$$CH_4 + \frac{1}{2}O_2 \rightarrow CO + 2H_2 + 35.7\,kJ.mol^{-1} \qquad [1.3]$$

The drawback of these two types of production is the release of carbon dioxide (CO_2), which is mainly responsible for greenhouse effect. A further drawback is the use of fossil energies as primary source.

1.3.1.2. *Water electrolysis*

The conversion of electric energy into hydrogen (and oxygen) thanks to water electrolysis is a long-known process, which was demonstrated by Faraday in 1820 and has been widely used since 1890. This solution remains the most interesting in terms of greenhouse gas emission if the electric energy used for decomposition does not emit CO_2.

The reactions taking place at the anode (negative electrode, [1.4]) and at the cathode (positive electrode, [1.5]) are the following:

$$H_2O \rightarrow \frac{1}{2}O_2 + 2H^+ + 2e^- \qquad [1.4]$$

$$2H^+ + 2e^- \rightarrow H_2 \qquad [1.5]$$

Water electrolysis is particularly well suited when high-purity hydrogen is required. Hydrogen production by electrolysis is currently three to four times more expensive than that by natural gas reforming. Nevertheless, the problem of costs can be overcome if the electricity used for electrolysis is derived from various renewable energy sources, such as wind or solar energy. Indeed, there are cases when this type of electricity can be produced while no other local demand or option of energy transfer to a neighboring region is required [AFH 18]. This electricity can then be perceived as free (being otherwise lost and dissipated). It could then be used for electrolysis, producing hydrogen that could afterward be stored for subsequent use.

1.3.1.3. *Direct production from biomass*

Biomass includes all the plants (wood, straw, etc.) that are renewed at the Earth's surface. It is a significant potential source of hydrogen. Indeed, gasification results in a $CO + H_2$ mixture. This needs to be purified in order to obtain hydrogen. This is an attractive solution, as the quantity of CO_2 emitted during the conversion of biomass into hydrogen is approximately equivalent to that resulting from photosynthesis: the environmental balance sheet is therefore zero.

1.3.1.4. *Hydrogen storage*

Hydrogen storage is crucial for the development of a chain with hydrogen-based energy vector. It is also essential for transportable applications and very interesting for applications with intermittent primary energy [CEA 18].

The increase in renewables' share of energy supply requires the implementation of energy storage systems (ESS). Renewable energy sources often provide intermittent power supply. Solar energy provides power depending on latitude, season, weather conditions and so on. As with solar energy, wind energy provides power that varies with wind speed.

Thus, the electric power generated from renewable energy sources cannot comply with the demand variation. However, a correlation must exist between offer and demand. For this purpose, a storage level is needed between the source and the load. Moreover, storage offers multiple advantages:

– environmental gain related to unlocking the large-scale deployment of decarbonized energies;

– capacity to bring centralized or decentralized responses to local or global constraints;

– non-dependence on fossil resources and long-term economic advantage, since the prices of resources and CO_2 are expected to increase.

The nature of this storage differs depending on the imposed specifications (amount of energy, power, overall external dimensions, cost, etc.). It is worth noting that direct storage of electric energy is not possible. First, electric energy must be converted into potential energy (mechanical, electrochemical, etc.) for storage, and then a second conversion is required to bring it back to its primary form [EXP 14].

The following are examples of electricity storage:

– electrochemical storage (accumulators);

– electrostatic storage (ultracapacitors);

– gravity storage, using compressed air or inertia (ETSP for energy transfer station by pumping; CAES for compressed air energy storage and inertia wheels);

– heat storage.

Here, the focus will be on chemical storage, which involves the use of electricity to obtain a chemical component or a molecule, which can then release energy by burning or via a fuel cell. Hydrogen and methanol are two such examples, with the emphasis on hydrogen, as it is associated with the fuel cell [EXP 14].

Hydrogen storage is a complex subject, as there are several more or less mature storage technologies with various efficiencies. How hydrogen is stored depends obviously on the application.

Transport applications require a storage volume, which is compatible with vehicle habitability, and a weight, which does not reduce vehicle performances. For stationary applications, constraints are less important. Table 1.1 provides the mass and volume storage densities [SOR 12, WUR 97].

Before presenting various possibilities of hydrogen storage for transport and stationary applications, let us recall its main advantages as follows:

– abundance on Earth;

– storage on demand, with no time loss;

– high energy density;

– no greenhouse gas;

– transport possibility;

– capability of being diluted into another gas (e.g. natural gas).

Storage form	Energy density		Density
	$kJ\ kg^{-1}$	$MJ\ m^{-3}$	$kg\ m^{-3}$
Hydrogen, gas (atmospheric pressure)	120,000	10	0.090
Hydrogen, gas (200 Bar)	120,000	1,900	15.9
Hydrogen, gas (300 Bar)	120,000	2,700	22.5
Hydrogen, liquid	120,000	8,700	71.9
Hydrogen in metallic hydrides	2,100	11,450	5,480
Natural gas	56,000	37.4	0.668

Table 1.1. *Energy density and density of hydrogen in various storage forms*

Compressed storage:

Nowadays, compressed storage is the most widespread storage technique. Standard cylinders have pressure ranging from 100 to 200 bar, and for automotive applications, the storage pressure ranges from 250 to 350 bar. There are also tanks pressurized at 700 bar, which enable the storage of more energy (> 3 kg H_2) and provide users with acceptable autonomy. The material used to construct these tanks depends on their application. For a stationary application, in which mass is less constraining, steel or aluminum is used, while for transport applications, composite fiber is a good choice. Moreover, the cost of these tanks depends on the quantity of carbon fiber used to reinforce the composite envelope. Ongoing research efforts aim to develop "low-cost" carbon fibers with the same specifications, particularly in terms of safety. Moreover, the thickness of the tanks should be maintained at reasonable values, in compliance with the norms for density [SOR 12, SAL 07].

Liquid storage:

Liquid hydrogen (LH_2) tanks can, in principle, store more hydrogen for a given volume than compressed hydrogen tanks, since the density of liquid hydrogen (0.070 kg/L) is higher than that of compressed hydrogen (0.039 kg/L) at 700 bar [CAR 81]. Nevertheless, the technological obstacles related to this type of storage are hydrogen evaporation, the energy needed to liquefy hydrogen and to maintain a temperature of −253.15°C and finally the cost of the tank. This process is not adapted to transport applications, but rather used for industrial applications.

Storage in metallic hydrides:

Some very porous materials enable hydrogen atoms to adhere at their surface. Metallic hydrides (nickel, titanium, magnesium alloy) have the ability to reversibly absorb and desorb hydrogen [CEA 18]. Storage in hydrides enables a density that is twice that of liquid hydrogen storage. Nevertheless, it can be achieved at the expense of weight, as the weight of the material in which hydrogen is inserted should be considered. Moreover, this technology requires good control of thermal phenomena during hydrogen loading and unloading. In fact, the absorption reaction is exothermic, whereas the desorption reaction is endothermic. This can be perceived as a safety guarantee, as hydrogen release is not possible unless there is a heat supply (between 75 and 125°C, depending on the material). Moreover, these tanks operate at low pressure (between 3 and 20 bar), which is a real advantage compared to other forms of storage in a transport application [BOT 07, SOR 12].

In summary, hydrogen is a promising energy vector, but there are still several obstacles to overcome before deploying a "hydrogen economy". Currently, there are technological, economic, security and social limitations. The most promising path is high-temperature water electrolysis. While electricity accounts for 80% of the production cost of hydrogen by electrolysis, the introduction of heat into the chemical reaction enables a reduction of the amount of electricity needed for the reaction. The advantage of this production technique involves a cost of less than 3 €/kg if decarbonized electricity is used. Although this is more expensive than the production method based on fossil energy (2 €/kg), electrolysis releases no greenhouse gases.

With the use of hydrogen in transport being the most promising in terms of reducing CO_2 emissions, the development of hydrogen must also go beyond the storage and fuel cell constraints. Mobile storage of hydrogen poses difficulties in terms of costs because of the decrease in efficiency related to its compression. Hence, the use of metallic hydrides seems a proper alternative for transport applications, since this storage is made at low pressure. Moreover, it is worth noting that for stationary applications, hydrogen can be distributed via existing networks of natural gas (up to 10% of H_2 in natural gas).

Finally, hydrogen faces a *problem of social acceptability*. Similar to LNG, hydrogen is highly flammable, and the fears of explosion risks should be countered by means of education, in order to avoid losing a large segment of users [UFE 14].

Hydrogen is an inexhaustible and "clean" energy resource. When coupled with a fuel cell (FC), hydrogen offers an interesting alternative, opening the way to an extraordinary technological and economic revolution. In the following section, fuel cell technology and its applications are presented.

1.4. Fuel cell and its applications

Global population is becoming increasingly sensitive to subjects related to future energies and the safety of our planet. This awareness has led us to find sustainable solutions for the most efficient possible energy transition. As described in the previous sections, there are various possibilities of replacing fossil resources and particularly oil resources by renewable energies that release little or no CO_2. Among them, hydrogen can be a "missing link", partly compensating for the major limitations represented by the lack of capacity to store electricity under satisfactory conditions [PÉR 06a]. Nowadays, private and public institutions are highly interested in this clean energy vector, as well as in fuel cells, which are highly efficient devices for energy conversion [COM 03, HIS 04]. Indeed, a fuel cell (FC) is an electrochemical energy converter that converts the chemical energy of a fuel into direct current (DC) [BAR 13].

1.4.1. *Brief history*

The fuel cell principle was described as early as 1839 by the English scientist Sir William Grove [GRO 39], who elaborated the idea that since it was possible to separate water into its components by means of electricity (electrolysis), the reverse reaction should be possible [BOU 07]. Following this discovery, the FC has remained a scientific curiosity for more than a century. During this period, Friedrich Wilhelm Ostwald (Nobel Prize laureate in 1909 for his works on catalysis) laid the theoretical foundation for FC operation. He mainly realized that internal combustion engines were limited by the Carnot cycle and that they were highly polluting, while the FCs that directly generate electricity were very efficient, silent and did not release pollutants. W.F. Ostwald predicted a technological revolution, as well as the fact that this transition would be very lengthy [OST 94].

It is only in 1939 that an English engineer, Francis T. Bacon, started working on a practical application that enabled him to characterize a 5 kW FC in 1952. However, the first practical applications were implemented in the American spaceflight programs at the early 1960s. General Electric developed the first proton-exchange membrane FCs used in Project Gemini. Further FCs were implemented in the Apollo project, which enabled the generation of electricity aboard a space shuttle, as well as the purification of water, which could be consumed by the astronauts.

This technology was very reliable, but had the disadvantages of complex implementation, high cost and relatively low power density. Consequently, FC technology was "once more" abandoned until the early 1990s.

Figure 1.3. *Fuel cell used in Apollo missions. For a color version of this figure, see www.iste.co.uk/jemei/cells.zip*

1.4.2. *Fuel cell and its revival*

Since the early 1990s, fuel cells have been drawing renewed interest, which seems to have further intensified over the past years. Among others, a particular manifestation of this new interest is a genuine willingness to protect the environment, problems associated with conventional sources of energy or even the necessity to develop new technology. The reduction in the emission of pollutants or greenhouse gases is an important work direction, especially in the transport sector, as it represents, on its own, more than a quarter of CO_2 emissions in France (Figure 1.4).

Hence, for several years, joint research actions have been conducted by transport material manufacturers, fuel cell manufacturers, contractors and research laboratories. In most such programs, proton-exchange membrane (PEM) fuel cells are at the basis of the system, being the only ones meeting the required dynamics, temperature, mass and volume power density criteria for transport applications. Moreover, they are also adapted for stationary applications (small-scale electricity generation) as well as for mobile applications. The works presented in this book rely essentially on this type of technology.

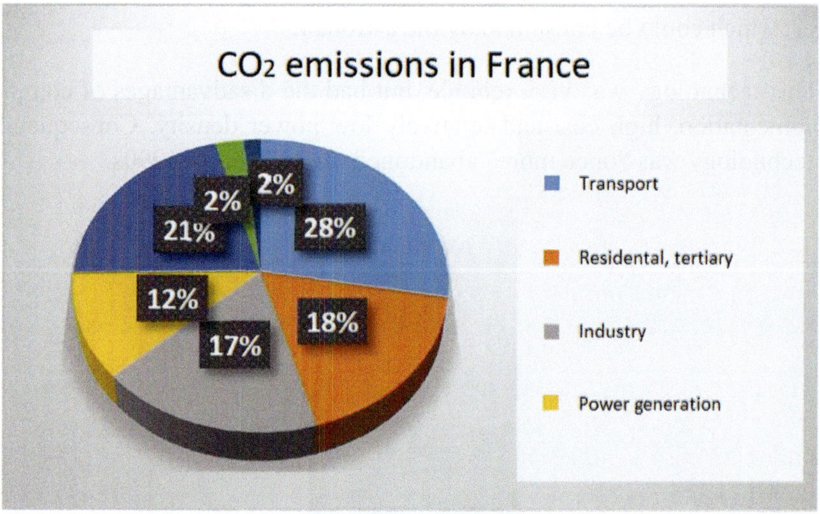

Figure 1.4. *CO_2 emission in France in 2012. For a color version of this figure, see www.iste.co.uk/jemei/cells.zip*

1.4.3. *Applications of fuel cells*

The fuel cell is progressively getting out of its exploratory phase. The use of this technology has been considered or experimented in many fields of activity. However, large-scale industrialization has not yet started, even though series productions have been announced, particularly in transport applications. Here the focus will be on applications that may emerge on the short to the medium-term. They refer essentially to transport and stationary applications. Table 1.2 provides the main characteristics in terms of power, fuel used, lifetime, operating mode, voltage level or cost.

	Automobile	**Stationary (primary source)**	**Stationary (emergency)**
Power	50–100 kW	1–10 kW & 200 kW	1–10 kW
Fuel	Hydrogen	Reformate	Hydrogen
Lifetime	5,000 h	> 40,000 h	< 2,000 h
Type of operation	Intermittent	Constant	Intermittent
Voltage	> 300 V	110/220 VAC	24 or 48 VDC or 110/220 VAC
Cost	< 45 €/kW	< 900 €/kW	< 4,500 €/kW

Table 1.2. *Characteristics of a FC system depending on the considered application [BAR 13]*

1.4.3.1. *Transport*

To develop sustainable mobility or sustainable transport, a review of the technologies employed by the currently used means of transportation is needed. Various sectors (road, railway, maritime and air transport) are concerned, but passenger and goods transportation is one of the main development directions.

Most car manufacturers have developed several prototypes of fuel cell vehicles (Figure 1.5). Some have even series production lines and started marketing their vehicles. It is the case of Hyundai and its ix35 or Toyota and its Mirai [KON 15]. Selling these vehicles to consumers is certainly a turning point in the automobile industry, as the idea of fuel cell vehicle introduction to the market is a dream dating back to the early 1990s. Indeed, the fuel cell has numerous environmental advantages. Apart from the contribution to reducing the greenhouse effect, fuel cell-powered vehicles, commonly known as zero emission vehicles (ZEVs), enable the improvement of urban life quality thanks to their low noise and reduced environmental pollution. However, many technological and economic obstacles still remain to be overcome before massive marketing of fuel cell vehicles. Reliability, cost and availability of hydrogen reload infrastructure can be immediately highlighted. Nevertheless, because of the joint efforts of various research institutes worldwide, as well as of several large industrial groups and car manufacturers, this is a time with the best perspectives for marketing development. In fact, the emergence of this new technology in the consumer car field will undoubtedly be achieved step by step in the near future.

For car applications, several possible configurations can be distinguished [BAR 13, RAJ 00]:

1) the FC is sized to provide the vehicle with the total power needed. Generally, a battery is needed to ensure the startup of the FC system. This configuration is only possible with onboard hydrogen use. Indeed, an onboard fuel reformer would not have the dynamics required for completing a transport mission;

2) the FC is sized to ensure supply of the mean power (constant power). Power peaks, particularly those required for accelerations will be provided by other ESS, with faster dynamics than those of electrochemical accumulators or ultracapacitors. Hybridization with accumulators enables the use of the FC in operating points, limiting its degradation. Moreover, ESS make it possible to recover the braking energy, which increases the overall efficiency of the vehicle, which is called parallel hybridization. Various architectures are detailed in Chapter 3;

3) the FC is sized to only recharge the accumulators. They provide the power needed for a vehicle to move. Here the architecture used is series hybridization;

4) the FC is used as an auxiliary power unit (APU) to provide electricity to the board network, powering various electric actuators (power steering, climatization, navigation, etc.). The APU is mainly used for heavy trucks or refrigerated vehicles, as it enables the supply of the reefer even when the heat engine is stopped.

Having listed the various possible configurations, we will now consider the specifications offered by fuel cells in terms of efficiency, emissions, cost, lifetime, cold start, integration or even recharging infrastructure [BAR 13, DOE 14].

The efficiency of a fuel cell system (which will be further detailed in Chapter 2) can reach 50% at its best operating point (corresponding to a hydrogen consumption of 60 g.kWh^{-1}). For an internal combustion engine (ICE), the efficiency for one gasoline supply reaches 34% (corresponding to a fuel consumption of 240 g.kWh^{-1}), while for a diesel supply, it is possible to reach an efficiency of 40% [KON 03]. It is worth noting that 1 g of hydrogen contains the same amount of energy as 2.73 g of gasoline (based on LHV – lower heating value). Nevertheless, it is not appropriate to compare the efficiencies of the FC system and ICE. In fact, their maximum values are not obtained at the same operating points. A FC reaches the highest efficiency when operating at average power [STO 03], whereas an ICE needs to reach maximum power in order to achieve the highest efficiency [MAS 03].

Figure 1.5. *Fuel cell vehicles (1. Toyota Mirai, 2. Hyundai ix35, 3. Daimler F-Cell, 4. Honda FCX-Clarity, 5. GM Equinox, Kia Borego FCEV). For a color version of this figure, see www.iste.co.uk/jemei/cells.zip*

Regarding the pollutants emitted by a FC, it can be immediately noted that a FC operating with onboard stored hydrogen releases no pollutants. On the contrary, if the primary energy is of fossil type (methanol or gasoline) which is reformed onboard the vehicle, then the global system displays the polluting emissions that will nevertheless remain below ICE emissions. Obviously, this is no longer valid if the pollutant analysis takes into account the global chain of energy production, commonly referred to as well-to-wheels. If hydrogen production (in a refinery, in a recharging station or onboard a vehicle) relies on fossil fuel, the resulting emissions, and particularly those of CO_2, have to be taken into account. Figure 1.6 provides a study of well-to-wheels, which takes into account the greenhouse gases produced during fuel fabrication processes, vehicle use and vehicle manufacturing [PEH 03]. It naturally appears that FC vehicles are less polluting than conventional vehicles using gasoline, diesel or methanol as fuel in an ICE.

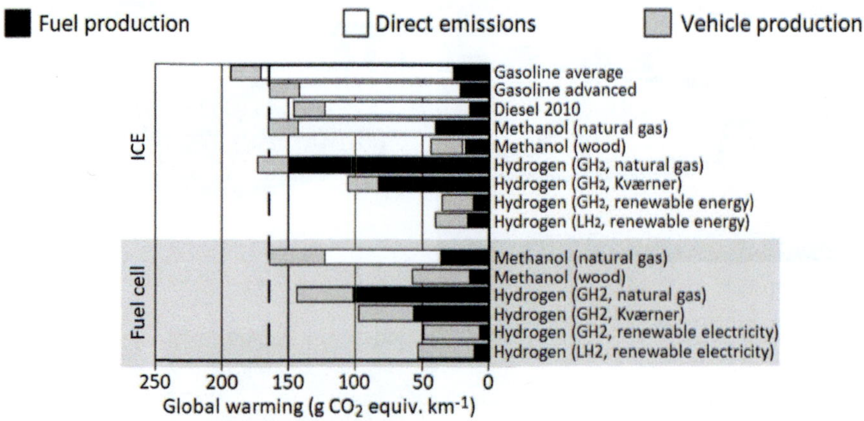

Figure 1.6. *Well-to-wheels life cycle analysis depending on fuel and on developed vehicle technology*

Nowadays, the ICEs of conventional vehicles are mass-produced in specialized manufacturing plants. This allows for much lower costs (30–45 €/kW) compared to the FC (220 €/kW for 1,000 systems/year). The FC remains an immature technology, and their production remains limited to prototypes or several hundred preseries units. It is evident that the cost of this technology can only be diminished by its large-scale development. This has been validated by most major car manufacturers. Indeed, by applying

ICE industrial fabrication processes to the FC, the 45€/kW figure could be reached if 500,000 FCs are manufactured per year [JAM 12].

To reduce this cost, many studies have been conducted on reducing platinum [ÇÖG 15, MRÓ 15] in the FC. This precious metal is used as a catalyst and, with ionomer membranes, it is one of the most expensive materials. Current FCs have 0.85 g/kW of platinum, which corresponds to a cost of 45 €/kW, which is a significant part in the overall cost. Acceptable objectives are of the order of 5 €/kW. Finally, for membranes, the current cost (45–65 €/kW) could be halved, provided mass production started.

Finally, the objective of 30–45 €/kW should be reached thanks to less expensive membranes, platinum reduction without performance alteration and mass industrialization of FCs.

The cost is obviously one of the obstacles on the mass development of FCs. Nevertheless, this technology also has a lifetime problem.

Today, conventional vehicles have a lifetime of approximately 10 years with an average mileage of 250,000 km. Given an average speed of 50 km/h, this corresponds to 5,000 h of operation. It is therefore the minimum lifetime that a FC must reach to be able to compete with ICE. Laboratory tests have shown that it is difficult to ensure more than 3,000 h of operation without significant degradation of performances. Better results can be obtained if the FC is directly connected to a stable rated operation. Chapters 3 and 4 explain how the lifetime of FCs can be improved either by hybridizing them with other energy sources or by using diagnosis and/or prognosis methods.

A further important point to consider in the case of a FC for transport applications is its cold start. Indeed, a conventional vehicle can operate in the temperature range of −40°C to + 50°C. For a FC, cold start is one of the technological obstacles, as below 0°C, the water contained in the FC membranes (see Chapter 2) freezes and blocks all electrochemical reaction. In the most unfavorable cases, freezing may result in FC mechanical breakage. There are various solutions to overcome this difficulty, such as the use of a cooling liquid that does not freeze (e.g. glycol), but this liquid can be corrosive and lead to significant FC degradations on average term. A protocol based on the purge of various FC circuits has enabled our team to start a FC at −20°C in less than 30 s with no external power contribution.

Despite many efforts made in recent years on the weight and volume of the FCs, the integration of a FC system on board of a vehicle remains an important factor to consider, as a FC vehicle must have the same habitability as a conventional one. Today, many FC manufacturers display interesting performances in terms of mass power density (1 kg/kW) and volume power density (1 l/kW). Nevertheless, the full FC system is logically heavier and bulky. Thus, it is difficult to have values below 2.5 kg/kW and 2.5 l/kW for mass and volume power densities, respectively. The average term objective is to obtain a mass power density of 1.54 kg/kW and a volume power density of 1.18 l /kW for the full FC system without hydrogen tank or electric motor [DOE 14].

Table 1.3 presents a list of objectives that will enable FC vehicles to compete with conventional vehicles.

Characteristics	***Units***	***State of the art (2012)***	***Objectives (2020)***
Efficiency at 25% of the rated power	%	59	60
Volume power density	W/l	400	850
Mass power density	W/kg	400	650
Cost (for 500,000 units/year)	€/kW$_e$	45	37
Cold start up to 50% of rated power			
From −20°C to ambient temperature	s	20	30
From +20°C to ambient temperature	s	< 10	5
Energy contribution for cold stop/start			
From −20°C to ambient temperature	MJ	7.5	5
From +20°C to ambient temperature	MJ	-	1
Lifetime over an automobile cycle	h	2,500	5,000

Table 1.3. *Objectives to be reached for the FC until 2020 [DOE 14]*

In the field of transport, public transportation – particularly urban transportation – is another application that could enable the introduction of FC technology in the short-term (Figure 1.7). Although the technological obstacles are similar to those for a light vehicle application, bus fleets could facilitate the introduction of hydrogen on the market of new fuels. Indeed, the use of bus fleets offers the major advantage of being able to recharge the hydrogen tanks of all the busses at the same location at the end of their route. A further advantage concerns the recharge and/or hydrogen production station, which is located on the same site. Moreover, the significant storage of hydrogen required (approximately 20 kg of H_2) for a FC installed power of 150 kW presents no particular drawback, since it is possible to place pressurized H_2 tanks, generally at 350 bar, on top of the bus. Moreover, the use of FC buses in towns would reduce the pollution output, as these ZEVs obviously release no pollutant. Moreover, these solutions provide 15% fuel savings compared to conventional buses with diesel-powered ICE [HOO 03]. Nevertheless, competitiveness with conventional buses and market introduction require an increase in FC lifetime. Indeed, classic buses can operate up to 6,000 h/year. Moreover, they are subjected to very constraining charging cycles, with very frequent stops/starts, which reduce the lifetime of the FCs. Even though the durability of FCs has significantly increased in recent years, it is still difficult to step up to the challenge proposed by the bus applications with current technologies. One of the paths that could lead to the improvement of FC durability for bus applications is the development of accelerated test procedures in order to reduce the time required to identify the various obstacles depending on driving profiles.

Commercial vehicles (Figure 1.7) are also part of high-potential FC applications. In recent times, lift trucks, transport vehicles in the airports, golf quadricycles or post office quadricycles have been in use. Most of these vehicles are electrically powered by lead-acid batteries, which require lengthy and frequent charging, the tool being rendered useless during recharging process. The use of a FC system to supply energy to these vehicles presents the advantage of a recharging time of only a few minutes per day and of maintaining the same power available between two charges, which is not the case with a traditional electric vehicle. Indeed, when the state of charge of batteries is too high, available power decreases. This may be extremely inconvenient for forklifts.

Figure 1.7. *Fuel cell vehicles pile (1. Bus, 2. Lift truck, 3. Airport transportation vehicle, 4. La Poste Quadricycle – MobyPost Project – FCLAB). For a color version of this figure, see www.iste.co.uk/jemei/cells.zip*

As noted in this section, there are many FC applications dedicated to transport. Nevertheless, there are still many obstacles to overcome. One of the main obstacles to the deployment of FC vehicles is the lack of hydrogen infrastructure. Indeed, there are no FC vehicles on our roads, as there are no recharging stations, and at the same time, there are no recharging stations because the number of FC vehicles is quasi-zero, therefore a vicious circle. Nevertheless, there are recharging stations worldwide, particularly in the United States, Japan and Germany. California has a network of recharging stations enabling the deployment of FC vehicles. In Japan, Toyota has joined Air Liquide in order to build 100 new hydrogen refueling stations (HRS) in 2015 in four large Japanese cities (Tokyo, Nagoya, Osaka and Fukuoka) and along the highways connecting these cities. Germany has committed to develop 400 stations before 2023, while France is expected to set up several dozen stations by 2020. This is an encouraging short-term perspective for Europe.

1.4.3.2. *Stationary*

Similar to transport applications, there are many stationary fuel cell systems that can be classified depending on application, power, installation, type of fuel or combined heat power (CHP) possibility or lack thereof.

The FCs for stationary applications can be used in industrial, residential or tertiary sectors and can also supplant or replace the power supplied by the electric grid.

The FC system has been designed to be employed according to various configurations [BAR 13]:

– in parallel with the network: the FC system enables the quasi-full supply of power required by the consumer, except for power peaks that are too significant and could diminish the lifetime of the FC or damage it. This configuration allows for a unidirectional flow of energy towards the consumer, if needed, but no power supplied by the FC can be re-injected into the network. This solution can do without battery (except for the start of the FC system);

– interconnected to the network: the flows of energy are in this case bidirectional. This enables the FC to operate at constant power (thus preserving its lifetime) and to re-inject power into the network when it is exceedingly generated by the FC;

– autonomous: the application is disconnected from the network. The FC system should be able to follow the charge variations imposed by the consumer;

– emergency power generator: the use of a FC system as emergency power generator requires rapid start and hybridization with batteries or ultracapacitors in order to provide power peaks. This application is advantageous compared to conventional emergency power generators for powers of several kilowatts and especially for energy supplies lasting more than 30 min. Finally, the ideal configuration is obtained when an electrolyzer, supplied by the network, generates hydrogen *in situ*.

These FCs have powers ranging from 1 kW for small applications to several megawatts for large-scale energy generation. The following classification of applications is proposed depending on the power range:

– 1–10 kW for single-family housing and mobile applications;

– 10–50 kW for small collective housing or small shops;

– 50–250 kW for small residential areas, hospitals, hotels or military bases;

– 250 kW–100 MW for centralized generation of energy.

There are various advantages depending on the application considered. For distributed power generation (> 250 kW), the use of FC technology enables the elimination of distribution and transmission losses, low polluting emissions, reliability increase and fewer problems related to the peak demand on electric networks.

A further characteristic of FCs is their high efficiency in CHP applications (simultaneous generation of heat and power), which may reach an efficiency above 80%. These systems are silent and generate no vibration. Moreover, they offer the possibility of using natural gas as fuel thanks to internal reforming, which is done by using the existing distribution network.

Although in terms of performance, specifications differ from those of transport applications, some technical obstacles are identical. For example, the overall cost of these fuel cell generators has to remain competitive and they need to offer the same possibilities as conventional generators.

Moreover, substantial progress needs to be accomplished over the lifetime of FC generators for the stationary applications to reach approximately 100,000 h of operation taking into account environmental factors such as ambient temperature (−40° to +40°C).

If PEM technologies are used for CHP, a further challenge is that their low-temperature operation limits the amount of thermal energy available. New technologies with higher operating temperatures (~200°C) must be developed. Hence, SOFC technologies are preferred for high-power CHP applications (> 100 kW). The operating temperature of an SOFC ranges from 800 to 1,000°C.

Indeed, SOFC is an increasingly mature technology for average-power CHP applications (> 100 kW). Moreover, many tests have proved an operation time of more than 25,000 h. It is preferable to use this type of FC on lengthy tests that do not require stops/starts. Indeed, thermal cycling should be avoided as the dilatation of materials in the cell core could generate leakages or even mechanical breakage. Moreover, the start time may prove very long (> 30 min). Hence, these two problems are major obstacles related to this technology. One of the solutions is to decrease the operating temperature to approximately 500°C. This technology is referred to as intermediate-temperature solid oxide fuel cell (IT-SOFC).

Finally, the last "family" of stationary applications is micro-CHP. This is applicable when powers remain below 10 kW. Low- or high-temperature FCs can be used. However, SOFCs may be preferred for this type of application, as they offer the possibility of using various fuels, such as natural gas, which is present in all housings. It should then be possible to replace conventional boilers in our housing with micro-CHP systems using natural gas that can supply the housing with electricity, heat and hot water. The challenges addressed by this type of system remain the cost and the increase in lifetime (Table 1.4). The technological obstacles are identical to those presented for average power applications.

Regardless of the stationary application, the FC system has a subsystem that enables the conversion of raw fuel into clean hydrogen or into a synthesis gas compatible with the chosen FC. Examples of raw fuels are natural gas or biogas (landfill gas, bio-methane, biodiesel, ethanol, etc.). All these fuels require various processing stages. Indeed, the passage of conventional fuels in a reformer generates a mix of hydrogen and carbon monoxide (CO). Or, the presence of CO at the anode of a PEMFC leads to blocking the activity of cell platinum catalysts and therefore to a degradation of its electrochemical performance. The operation of a PEMFC using gasoline or gas-oil requires the implementation of mechanisms between the reformer (800 and 1000°C) and the cell, enabling the elimination of CO [MUL 09]. Two reactions through which gas is converted into water then follow: the first reaction at about 400°C and the second one at about 200°C. Each of these two exothermic reactions converts between 80 and 95% of CO into CO_2. The last stage is a preferential oxidation catalytic reaction, also exothermic, taking place at about 150°C, which enables a decrease in the CO conversion rate from less than 1% to several dozens of ppm. Depending on the origin of the primary fuel as well as on the geographical region, they may contain a certain amount of sulfur. For a SOFC, a reversible poisoning of the anode takes place from 1 ppm sulfur at an operating temperature of 1000°C to 50 ppb at an operating temperature of 750°C.

Thus, the fuel processing subsystem can contain the following:

– desulfurization unit, in which preprocessing of primary fuel takes place in view of a maximum reduction of sulfur rate;

– reformer, in which thermal cracking of hydrocarbons takes place at high temperature;

– water-gas-shift, which enables the reduction of CO and the increase of hydrogen content in gas flow;

– separators that increase the quality of hydrogen.

The most important obstacles at the level of the fuel conversion subsystem are lifetime, cost, tolerance to impurities, flexibility of the system to accept various fuels and cold start. Efficiency improvement and cost reduction are nowadays a necessity. Improvement paths are directed towards better physical integration of the elements subjected to high temperatures by developing multifunction catalysts in order to facilitate various reactions in the same reactor. This could reduce costs and improve efficiencies by reducing the losses due to various reactors (reformer, desulfurization unit, water-gas-shift, etc.) and by reducing the number of sensors or even the control laws. Finally, the system is simplified and the costs are reduced thanks to the elimination of various components.

Characteristics	***State of the art (2012)***	***Objectives (2020)***
Electric efficiency at rated power	34–40%	> 45%
Energy (CHP) efficiency	80–90%	90%
System cost 2 kW	1,100 €/kW	900 €/kW
System cost 5 kW	2,100–3,700 €/kW	1,400 €/kW
System cost 10 kW	1,750 €/kW	1,550 €/kW
Response time (from 10 to 90% of P_{nom}*)*	5 min.	2 min.
Run-up time at ambient temperature	< 30 min.	20 min.
Degradation	< 2%, 1,000 h	0.3%, 1,000 h
Lifetime	12,000 h	60,000 h

Table 1.4. *Objectives to be met by FCs in stationary applications by 2020 [BAR 13]*

1.4.3.3. *Other application*

Another application that is worth presenting and can be classified as either transport or stationary application is the auxiliary power unit, commonly designated by the acronym APU. These are systems that enable the supply of electric energy for built-in accessories in transport applications (heavy trucks, refrigerated trucks, boats, camping-cars, locomotives, airplanes, etc.) independently of the traction chain. Within stationary applications, APUs can also be used for supplying electric energy to isolated sites. The focus will be here on transport applications.

The market of APUs for heavy trucks appears as the most promising for APUs marketing in the near future. Today, FC APUs offer an interesting solution to the *idling* practice in heavy trucks. For a long time, the United States has seen an increased public and political awareness on the necessity to stop the common practice of running the diesel engine of heavy trucks when they are stopped in a rest area, parking, on the roadside or at a loading dock. The driver who is resting in the cab has the advantage of enjoying the comfort functions when the vehicle is stopped. Typically, heavy trucks resorting to *idling* are semi-trailers equipped with a sleeping cab (Figure 1.8) and used for goods transportation over daily distances of above 800 km. According to a survey conducted by Lutsey *et al.* [LUT 04], the main reason for heavy truck drivers to run their engine while not in motion is the need to regulate the cab temperature by heating or climatization, depending on the season. The second motivation would be the need to use the electric accessories that are present in most American heavy trucks equipped with a sleeping cab. The last reason is the desire to maintain the engine and its oil at adequate temperature in order to avoid restart problems during winter.

Figure 1.8. *APUs for heavy trucks – example of American heavy truck (Class-8 Heavy Duty Truck) with typical resort to idling. For a color version of this figure, see www.iste.co.uk/jemei/cells.zip*

There are multiple problems related to this practice. On the one hand, the impressive diesel engines of heavy trucks are designed and optimized to have optimal efficiency when they tow goods trailers at speeds stabilized on roads and highways. When idled with the vehicle stopped, and consequently over a use range for which they are not adapted, their efficiency becomes extremely low. According to Brodrick *et al.* [BRO 02], this efficiency sets between 9 and 11% only for the engine, consequently driving a significant fuel overconsumption. Moreover, several studies conducted by the *Center for Transportation Research* of *Argonne National Laboratory* (ANL) [GAI 06, LUT 05] show that out of 2.5 million heavy trucks inventoried in the United States, 458,000 would let their engine run while idle when stopped during the night in rest areas, parking or on roadsides.

The use of APU in heavy trucks reduces greenhouse gas emissions. Taking into account only the 500,000 heavy trucks classes 7 and 8 in the United States, the emissions released during *idling* periods amount to 10.9 million tons of CO_2 and 190,000 tons of NO_X per year [MUL 09].

The challenges of these applications are the cost, the operation in transient state (especially for APU SOFC) and the necessity to use fuel present in the vehicle (generally diesel) without adding specific subsystems (e.g. water circuit for the reformer). Furthermore, it is difficult to take into account the vibrations and shocks of the vehicle. Moreover, the APU must be optimized in terms of weight and volume in order to be integrated into the dedicated space in the vehicle. For this purpose, this technology should have increased power densities.

As a final point of this overview, it is worth mentioning the mobile applications with power below 250 W. This power range addresses directly to battery chargers, mobile phones, portable computers, tablet PCs and so on. Nevertheless, it is important to increase the energy density with respect to the current technology in order to fully meet the expectations of users in terms of autonomy. Further challenges are the cost, efficiency and the necessity to reduce the overall dimensions of the FC surrounding system, which is required for its proper operation.

1.5. Conclusion

In this chapter, we have highlighted the high potential of FC technology through its various and multiple applications. Fuel cell development is an attractive and promising alternative, as it would contribute to overcoming the energy dependency on oil reserves, while proposing little or non-polluting solutions. There is still room for progress and several obstacles need to be overcome for the deployment of a fuel cell market. Problems related to cost, efficiency, lifetime, integration or clean hydrogen production have been frequently mentioned, but the advantages offered by this technology should also be highlighted. The most important among them are as follows:

– High energy efficiency: regardless of the application, FC efficiency is above that of internal combustion engines or of systems for conventional energy conversion.

– Less or non-polluting: the use of a hydrogen-fueled FC releases no polluting emissions. Nevertheless, since hydrogen is often obtained from hydrocarbons, some CO_2 and/or CO emissions are generated. The amounts emitted remain however below those produced by classic energy conversion technologies.

– Cost reduction: the costs involved in fuel cell fabrication will drastically decrease as soon as mass market emerges. Indeed, FCs consist of a stack of identical cells. The manufacturing process will therefore be very easy to automate and execute. Some materials that are essential for the proper operation of FCs (sulfonated fluoropolymer) nevertheless remain expensive. It is therefore necessary to continue the development of mass manufacturing techniques in order to reduce the costs of FC components and the surrounding system.

– Lifetime: FC lifetime is still a scientific obstacle to overcome. Nevertheless, FCs have no rotating parts and should in time have very long lifetimes. For transport applications, the objective is about 8,000 h, while stationary applications should reach 80,000 h.

– Silent: FCs are silent, which renders them interesting for certain applications, such as military applications. Moreover, FC technology will improve the life quality in numerous situations (vehicles in down-town areas, handling equipment in the factories, etc.).

In terms of design, the fuel cell is a relatively simple tool. For its best operation, implementation requires nevertheless a system surrounding the cell core: this is referred to as a fuel cell system. A fuel cell system is a complex device in which numerous phenomena of various natures interact and need to be controlled. These explanations are essentially the object of Chapter 2.

2

From FC to System

2.1. Introduction

The fuel cell (FC) principle was described as early as 1839 by the English scientist Sir William Grove. Since then, the interest for this technology has gone through many unpredictable circumstances. Nevertheless, over the last 30 years, this technology has been subject to a rediscovery process. In fact, the evolution of the atmosphere has become a major problem, due to the constant increase in the concentrations of polluting emissions. In recent years, manufacturers, fuel cell producers, contractors and research laboratories have combined their research efforts to address this problem. Most of these programs focus on FCs with proton exchange membrane (PEM), on which the system relies, as they meet the required dynamic criteria (particularly for transport applications), as well as the temperature and mass and volume power density criteria. Nevertheless, another technology can satisfy the demands of a transport application, if it is used as an auxiliary power unit (APU), namely the solid oxide fuel cell (SOFC).

Regardless of the technology used, it is important to be able to control, implement, characterize and model the FC generator.

Chapter 1 showed that fuel cells had their place in the energy transition, either for transport or stationary applications. Nevertheless, the system surrounding the generator should be able to make it operate at its best parameters. Thus, it is important to consider in detail all the auxiliaries that constitute the "fuel cell system".

In the first part of this chapter, the various existing fuel cell technologies will be defined, with a focus on those that are best suited to transport and stationary applications. As the second part will show, system approach differs depending on the FC technology and the retained application. One of the key points of the FC system is its experimental implementation, which will be described in the third part of this chapter. Finally, the limitations of FC systems and the means to overcome them will be presented.

2.2. Fuel cell technologies for transport and stationary applications

2.2.1. *Presentation of various technologies*

A fuel cell is a generator that directly produces electricity through an electrochemical reaction. The elementary cell of a fuel cell has two electrodes, a positive one (cathode) and a negative one (anode). Electrochemical reactions take place in these electrodes. Thanks to the presence of the catalyst (e.g. platinum) at the electrode–electrolyte interface, the fuel (hydrogen, methanol, etc.) is oxidized at the anode and the oxidizer (oxygen) is reduced at the cathode. These two electrodes are separated by an electrolyte, which can be solid or liquid. In the case of a PEMFC, the electrolyte is a proton H^+-conducting polymer membrane, which enables the ionic transfer between the anode and the cathode. It must be at the same time a good ion conductor and an electron insulator. The electrons generated are evacuated through the external electric circuit.

The membrane electrode assembly (MEA) includes the membrane, the catalyst and gas diffusion layers (GDLs). To avoid gas leakage, MEAs are surrounded by specific seal gaskets. These gases are brought to the level of the assembly by bipolar plates. This assembly then constitutes an elementary cell (Figure 2.1), which needs to be duplicated in order to get a stack, called the fuel cell.

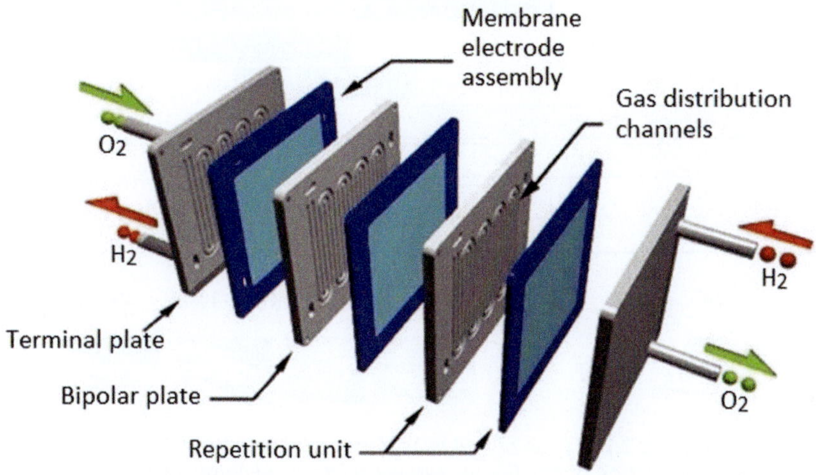

Figure 2.1. *Schematic diagram of a fuel cell [CEA 07]. For a color version of this figure, see www.iste.co.uk/jemei/cells.zip*

There are six types of fuel cells (Table 2.1), classified according to their electrolyte or operating temperature (Figures 2.2 and 2.3). Among them, two are commonly used in transport and stationary applications. The PEMFC is widely used in applied research fields, due to its advantages; similarly, the SOFC is often used in stationary applications. The next section details the operation of these two fuel cell types.

FC type	*Temperature*	*Electrolyte*
Phosphoric acid fuel cell (PAFC)	150–200°C	H_3PO_4
Polymer electrolyte membrane fuel Cell (PEMFC)	50–100°C	Polymer membrane
Direct methanol fuel cell (DMFC)	50–100°C	Polymer membrane
Alkaline fuel cell (AFC)	25–75°C, 100–250°C	KOH
Molten carbonate fuel cell (MCFC)	600–700°C	Molten carbonate salt (CO^{3-})
Solid oxide fuel cell (SOFC)	500–1,000°C	Ceramics (zircon oxide)

Table 2.1. *Types of fuel cells depending on operating temperature*

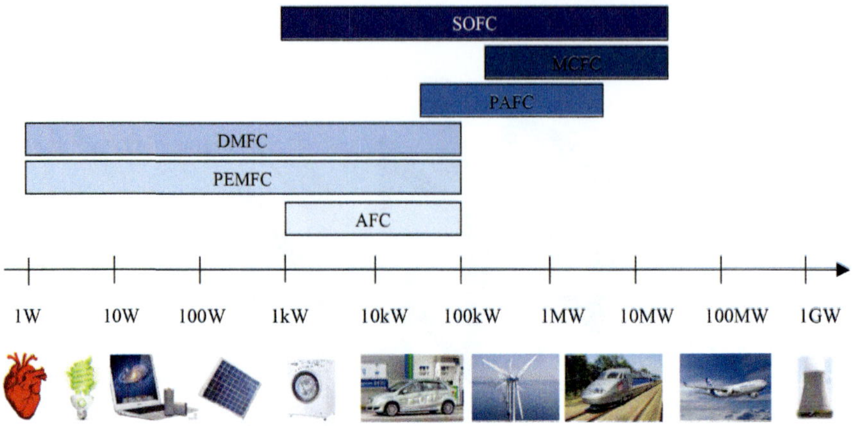

Figure 2.2. *Types of fuel cells depending on power. For a color version of this figure, see www.iste.co.uk/jemei/cells.zip*

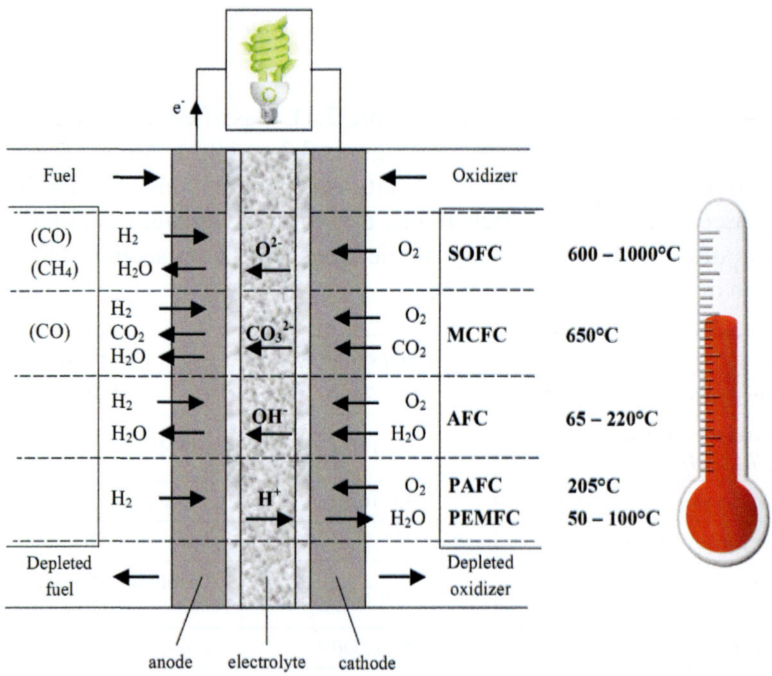

Figure 2.3. *Types of fuel cells depending on electrolyte operating temperature. For a color version of this figure, see www.iste.co.uk/jemei/cells.zip*

2.2.2. *Operating principle*

2.2.2.1. *PEMFCs*

PEMFCs – proton exchange membrane fuel cells – provide high power density and offer volume and mass packing density below those of other fuel cell technologies. PEMFCs use a very thin proton-conducting polymer membrane (< 50 µm), which constitutes the electrolyte. Their electrodes are made of porous carbon containing a platinum alloy catalyst required for the separation of hydrogen electrons and protons. It is worth noting that the platinum catalyst is extremely sensitive to carbon monoxide, which involves the use of an additional reactor to reduce the carbon monoxide content in the fuel, if hydrogen results from hydrocarbon reforming. Moreover, the presence of sulfur drives the rapid and irreversible degradation of performances [PÉR 06b]. Their operation requires supply of anode with hydrogen and of cathode with oxygen (or oxygen in the air), as well as humidification of input gases. Typically, they are fed with pure hydrogen from a (low- or high-pressure) storing tank.

PEMFCs operate at low temperatures (between 50°C and 100°C), which enables rapid start (no preheating) and a more substantial lifetime (less constraints on materials). This technology is considered the most "promising" for transport applications due to the solid electrolyte, high efficiency, high power density, rapid start and fast response to charge variations.

Figure 2.3 summarizes the operation of PEMFCs. The fuel is generally hydrogen, which combines with oxygen to yield water according to the chemical reaction below (equation [2.1]). This relation corresponds to that of hydrogen combustion. The energy of this reaction is released as heat and electricity. Combustion is facilitated by the platinum (Pt)-based catalyst:

$$H_2 + \frac{1}{2} O_2 \rightarrow H_2O \qquad [2.1]$$

To quantify the energy released during the reaction, the energy released during the combustion of reactants is generally used as a reference. This energy is called enthalpy ΔH, which depends on the state of reactants before and after the reaction. The enthalpy of reaction is expressed as

$\Delta H_R = \Delta H_{products} - \Delta H_{reactants}$. There are two possibilities for the combustion of hydrogen yielding water [FRI 03].

The first one is water generation in the form of vapor (equation [2.2]):

$$H_2 + \frac{1}{2}O_2 \rightarrow H_2O_{vapor} \qquad [2.2]$$

$$\Delta H = -241.83\, KJmol^{-1}$$

If water condenses after combustion, then the value of enthalpy changes (equation [2.3]):

$$H_2 + \frac{1}{2}O_2 \rightarrow H_2O_{liquid} \qquad [2.3]$$

$$\Delta H = -285.84\, KJmol^{-1}$$

The difference between the two values corresponds to water evaporation enthalpy or latent heat. The negative value of ΔH explains heat generation ($\Delta H < 0$) during the reaction.

In the specific case of PEMFC, the two half-reactions can be written as:

– at the anode: $2H_2 \rightarrow 4H^+ + 4e^-$ [2.4]

– at the cathode: $O_2 + 4H^+ + 4e^- \rightarrow 2H_2O$ [2.5]

Protons are carried through the electrolyte by the water molecules it contains. Thanks to its low operating temperature, water contained in the membrane in liquid form is preserved. Membrane dry-out induces an increase in proton resistance and thus a decrease in the electric performance of the fuel cell. Therefore, careful water management is essential.

2.2.2.2. *SOFCs*

SOFCs are electrochemical converters whose main characteristic is very high operating temperature, generally ranging between 500 and 1,000°C, which is essential for reaching proper ionic conductivity of the ceramic solid electrolyte. The other characteristic of SOFCs is the use of fully solid electrolyte and electrode materials, which consequently renders the fuel cell concept simpler than in other fuel cell technologies, such as PAFCs and MCFCs. The main advantage of SOFCs is their ability to use carbon monoxide as reactant at the anode (unlike PEMFCs, for which CO is poison). They also tolerate traces of sulfur. It is thus possible to use conventional fuels, such as gasoline, gasoil or natural gas, without the need to develop a "hydrogen infrastructure". SOFCs can thus find short-term market opportunities. Moreover, SOFCs make high-quality heat available for combined heat and power generation applications (vehicle or home heating).

The electrolyte layer in a SOFC is dense and gas-tight. It thus enables the separation of fuel flows at the anode and of oxidizer flows at the cathode. The anode and cathode layers are porous, which facilitate the diffusion of gaseous reactants and reaction products. Apart from separating gaseous flows, the electrolyte facilitates the movement of oxygen ions O^{2-} from the cathode to the anode. In the triple-contact area of the cathode, first, oxygen molecules O_2 dissociate into oxygen ions O^{2-} by accepting electrons from the anode (equation [2.6]). Then, oxygen ions O^{2-} migrate through the electrolyte towards the triple-contact areas of the anode (point between the gas, electrolyte and catalyst). In a fuel cell supplied with pure hydrogen, oxygen ions O^{2-} recombine with hydrogen H_2 to form water vapor (H_2O) (equation [2.7]). The released electrons are then carried to the cathode by the external electric circuit, thus closing the electric circuit loop. A current is then generated when a charge is added to the external circuit (Figure 2.3). The equations involved are as follows:

Reaction at the cathode: $1/2\ O_2 + 2\,e^- \longrightarrow O^{2-}$ [2.6]

Reaction at the anode: $H_2 + O^{2-} \longrightarrow H_2O + 2\,e^-$ [2.7]

Overall reaction: $1/2\ O_2 + H_2 \longrightarrow H_2O$ [2.8]

In SOFCs, carbon monoxide (CO) can be used, similarly to hydrogen, as an oxidizer at the anode. It is then possible to directly send the mixture of hydrogen and carbon monoxide typically resulting from a hydrocarbon reformer to the SOFC, without an intermediate stage of CO extraction. Classically, the oxidation of carbon monoxide at the anode is presented as an electrochemical reaction at the anode of SOFCs [LAR 03, SIN 03]:

$$CO + O^{2-} \longrightarrow CO_2 + 2\,e^- \quad [2.9]$$

In reality, the reaction presented in (equation [2.9]) is part of a very small minority in SOFCs [HOL 99, MAT 00, NET 04]. The *water-gas shift* reaction is predominant, which involves the reaction of carbon monoxide with water vapor to form the product given in (equation [2.10]):

$$CO + H_2O \longrightarrow CO_2 + H_2 \qquad [2.10]$$

The hydrogen thus formed by the reaction of gas and water is then oxidized according to reaction [2.7].

2.2.2.3. *Other technologies*

AFCs are one of the first fuel cell technologies developed, which have been widely used in American space programs (Apollo and Space Shuttle) since the 1960s, in order to generate electricity aboard space shuttles. This technology uses potassium hydroxide (KOH) solution as the electrolyte and accepts several materials as catalysts (nickel, silver and all types of noble metals) at the anode or cathode. Its operating temperature varies from 65 to 220°C, depending on KOH concentration in the electrolyte. For high temperatures (> 200°C), the concentration is about 85%, and for the lowest temperatures (< 120°C), it varies between 35 and 50%. AFCs show excellent performances, and their efficiency can reach 60%. Nevertheless, one of the major impediments of this application is their intolerance to carbon dioxide (CO_2), which is equally present in the fuel and in the oxidizer.

PAFCs use an electrolyte composed of phosphoric acid contained in a porous silicon and Teflon matrix. Anode and cathode catalysts are made of platinum, whose operating temperature varies between 150 and 220°C. This technology is considered as the first generation of modern fuel cells. It is the most mature and the first to have been marketed. PAFCs are generally used for high-power stationary applications (> 200 kW) as their lower energy density renders them massive.

MCFCs have an electrolyte composed of alkali carbonates retained in a ceramic matrix. Their operating temperature ranges between 600 and 700°C. Their high operating temperatures allow for the use of non-precious metals as catalysts at the anode and the cathode. This quite mature technology is specifically used for high-power stationary applications.

DMFCs can be classified as another FC technology. They are nevertheless very similar to PEMFCs, except that they use methanol instead of hydrogen.

2.2.3. *Comparison of technologies*

In the previous section, we summarized the various types of existing FCs, with a focus on PEMFCs and SOFCs. A comparison of these technologies is proposed here, aimed at highlighting the best-suited solutions, depending on the intended application.

PEMFCs operate at a temperature of about 75°C. This low temperature enables rapid start, as the "preheating" time is diminished. This environment becomes less aggressive for FC components, which results in a longer lifetime. The cost is nevertheless increased due to the noble metal-based catalyst (generally platinum) required for the separation of electrons and protons of the dihydrogen molecule. Furthermore, the platinum catalyst is extremely sensitive to poisoning with carbon monoxide. If the hydrogen used for the supply of the PEMFC results from hydrocarbon reforming, a reactor will be needed for the reduction of CO content in the fuel. Obviously, the cost of this process adds up to the overall application cost.

PEMFCs are generally used for transport and stationary applications. Thanks to their rapid start, power density, density and solid electrolyte (for safety issues), PEMFCs seem to be the most promising technology for transport applications.

SOFCs operate at very high temperatures (about 800°C), which enables the use of less expensive catalysts. Moreover, this renders the internal reforming of various fuels possible, reducing the cost of the overall system, since the addition of an external reformer is no longer needed. SOFCs are the most sulfur-resistant technologies. They can tolerate significantly higher amounts of sulfur (up to several tens of ppm) than other technologies. Moreover, they tolerate carbon monoxide, which can be used as fuel. Consequently, natural gas, biogas or gases resulting from coal can be used as fuels.

High temperature is nevertheless associated with a certain number of drawbacks. The first drawback is the very long start time. System preheating should be carefully conducted, in order to avoid thermal shock on the various materials composing the FC and the system. This increase in temperature takes several tens of minutes. Moreover, to protect the surroundings and especially the users in the case of a transport application, high-performance heat insulation is required. Finally, the very high temperature involves very strong thermal mechanical constraints on the materials. Hence, when the system passes from ambient temperature to its operating temperature, the stop/start operations should be limited with this technology. High temperature results in more rapid degradation and therefore a shorter lifetime. The development of low-cost materials with longer lifetime at high temperatures is a technical challenge, which should be overcome to increase the performance of this technology.

AFCs operate at approximately 60°C. The new AFC technologies use a polymer membrane, which brings them closer to PEMFCs. The difference is that their membrane is alkaline, while that of PEMFCs is acidic. The main drawback, as well as a differentiating characteristic, is that AFCs can be easily poisoned by CO_2. Even a minimum content of it in the air can affect the FC operation. The air and hydrogen supplying the FC should therefore be purified, which renders the system more expensive. Lifetime is obviously reduced in the case of CO_2 poisoning.

PAFCs have the characteristic of exhibiting lower performances than other technologies at a given weight and volume. They are generally very heavy, bulky and expensive, as the catalyst requires more platinum than other FCs.

The operating temperature of *MCFCs* is approximately 650°C, which leads to the set of drawbacks listed for SOFCs. The high temperature and the use of a corrosive electrolyte accelerate corrosion and the breakage of components, reducing its lifetime. It is nevertheless worth noting that the overall efficiency of coupling this FC with a turbine can reach 85%.

Table 2.2 proposes a comparative summary of various technologies, with their advantages and drawbacks. It appears that PEMFCs and SOFCs are best placed as FC technologies to be used in short-term applications.

The notion of system has been introduced in this comparative section. Indeed, a certain number of auxiliary circuits are required for proper operation of the FC. It should thus be noted that, regardless of the technology used, various peripherals are required. When using a FC, the system aspect should therefore be considered. There are still many obstacles when conducting a system approach, and one of the main such obstacles is the efficiency of the overall system.

FC type	*Temp. (°C)*	*Stack power*	*Applications*	*Advantages*	*Drawbacks*
PEM	< 120°C	< 1 kW–100 kW	– Emergency supply – Nomadic supply – Decentralized power generation – Transport	– Solid electrolyte – Low temperature – Rapid start	– Expensive catalyst – Sensitive to impurities in the fuel
AFC	< 100°C	1–100 kW	– Military – Space – Emergency supply – Transport	– Low cost of components – Low temperature – Rapid start	– Sensitive to CO_2 in the fuel and in the air – Control of aqueous electrolyte
PAFC	150–200°C	10–400 kW	– Decentralized power generation	– Suited for µCHP – Intolerance to impurities in the fuel	– Expensive catalysts – Long start time – Sensitive to sulfur
MCFC	600–700°C	300 kW–3 MW	– Public utilities – Decentralized power generation	– High efficiency – Choice of various possible fuels – Suited for µCHP – Gas turbine compatible	– Corrosion due to high temperature and breakage of cell components – Long start time – Low power density
SOFC	500–1,000°C	1 kW–2 MW	– Auxiliary power unit (APU) – Public utilities – Decentralized power generation	– High efficiency – Choice of various possible fuels – Solid electrolyte – Suited for µCHP – Gas turbine compatible	– Corrosion due to high temperature and breakage of cell components – Long start time – Limited number of stops/starts

Table 2.2. *Comparison of various technologies*

2.3. System approach

Optimization of the overall energy efficiency of the FC system particularly involves the study of the auxiliaries and requires a system approach. Our team has conducted several studies on FC auxiliaries (humidifier, compressor, power converter, etc.) and systems (search for optimal FC generator architecture, design and realization of APU sources based on FCs and ultracapacitors, modeling and tests of APU SOFC source, etc.). In this section, we will first focus on the constituent elements of a FC system and then on the system architecture depending on the retained FC technology.

2.3.1. *Auxiliary circuits*

A fuel cell requires a large number of auxiliaries that are essential to its proper operation. A FC system is generally composed of the following circuits:

– fuel cell stack;

– fuel supply circuit;

– oxidizer supply circuit;

– cooling circuit;

– electric circuit;

– control system.

There are multiple scientific and technological challenges to overcome in relation to auxiliaries. Before identifying them, the role of each of the circuits comprising the FC system should be defined.

Fuel supply circuit:

This circuit must allow fuel supply of the fuel cell anode under very precise conditions of temperature, pressure, flow rate, composition and humidification, depending on the technology used. Regardless of its form, the fuel must be stored in a tank.

Oxidizer supply circuit:

The FC cathode compartment must be supplied with oxidizer under the same conditions as the gas sent to the anode, depending on the retained technology. The most widely used oxidizer is oxygen in the air, which eliminates the need for a storage level. Nevertheless, in certain applications (particularly in an anaerobic environment), pure oxygen needs to be used and therefore stored.

Fuel cell performances are improved by the increase in reactive gas pressure. Hence, a significant number of FC systems are composed of an air compressor, which enables the pressure to increase by a ratio ranging from 2 to 4, relative to the atmospheric pressure.

Cooling circuit:

Apart from electricity generation, thermal power generation should be considered in a fuel cell. The main role of this circuit is to maintain the fuel cell stack at its rated operating temperature and to evacuate the calories generated during an electrochemical reaction. Depending on the technology used, this circuit may also enable the use of the resulting heat.

Electric circuit:

The electric circuit must supply the load connected to the fuel cell as efficiently as possible. The FC generates direct current (DC), whereas many applications operate with alternative current (AC). A converter is therefore needed for the transformation and adaptation of the current supplied by the generator to supply the electric charge, as well as to supply the set of auxiliaries of the FC system. The main difficulty in using DC/DC converters is that the generator operates at low voltage and high current. In general, this involves significant sizing of components. Moreover, the ripples generated by the converter at the FC output (high-frequency commutations of the switchers) must be filtered to avoid the degradation of the FC.

Control/command system:

The monitoring system is a central element of the FC system, enabling its optimal operation. Indeed, the various auxiliary circuits are controlled and commanded so that the fuel cell operates within a range that ensures the security of both material and personnel, as well as reaches optimal energy efficiency.

Depending on the technology used, FC systems feature other auxiliary circuits. An example is the exhaust gas processing circuit for high-temperature FCs or the energy storage modules that are often used in FC systems.

Although FC cores and their auxiliaries have undergone significant progress, it turns out that the overall energy efficiency can still be improved by using dedicated components. Indeed, net efficiency is highly dependent on the power used by the auxiliaries, which is deducted from the gross power supplied by the FC, as expressed in the following equation:

$$\eta_{net} = \frac{P_{FC} - P_{aux}}{P_{chemical}} = \frac{V_{FC} I_{FC} - P_{aux}}{-\dot{\eta}_{H_2} \Delta H_f \, n} \qquad [2.11]$$

where ΔH_f is the enthalpy of formation of one mole of water, which also corresponds to the quantity of heat released by the full combustion of one mole of hydrogen, $\dot{\eta}_{H_2}$ is the molar flow rate of hydrogen consumed by one cell to provide the expected electric power, V_{FC} is the voltage across the FC, I_{FC} is the current supplied by the FC, P_{aux} is the power consumed by the set of auxiliaries and *n* is the number of cells of the FC.

A further way to improve the overall energy efficiency is to optimize the system architecture according to the retained FC technology. These subjects are dealt with in the next section.

2.3.2. *System architecture*

The fuel cell system involves a set of auxiliary circuits whose interaction ensures optimal operation of the generator. Depending on the retained technology, the minimum number of auxiliary circuits is five (oxidizer, fuel, cooling, electric control and monitoring). Humidification circuits (PEMFC) or exhaust gas reprocessing circuits (high-temperature PEMFC and FC) can be added to the five basic circuits. Figure 2.4 shows a FC system.

The strong coupling between these various circuits renders the system multi-physical (electric, fluidic, thermal, etc.) and multi-scale (time and space), which requires close control of each auxiliary, as well as overall monitoring of the system.

The system control/command strongly depends on the developed architecture. The latter differs depending on the retained FC technology as well as on the intended application (transport, stationary, portable, etc.). An important preoccupation is the integration of this system. The constraints in terms of mass and volume are completely different if a FC system is installed in a car or in a building.

The most significant difference between a stationary application and an automobile application lies in the architecture of the electric conditioning. Indeed, a transport-dedicated system must be fully autonomous, while a stationary-oriented system can operate autonomously, as well as in parallel with an existing electric network or as emergency supply.

The autonomous system must be coupled with another energy source (auxiliary source), such as ultracapacitors or (storage) batteries, to supplement the FC system during power peak demands, which should be avoided for FC systems, as well as to supply the energy required for the system's start. Architectures and energy management laws of such systems are the object of Chapter 3.

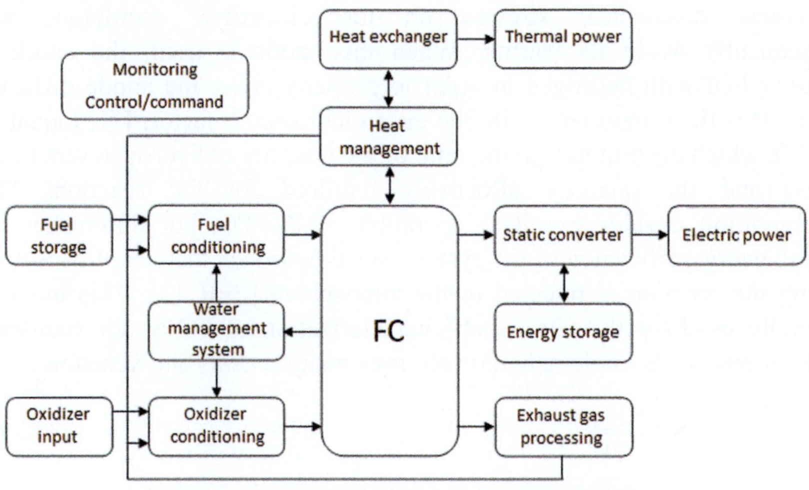

Figure 2.4. *Architecture of a fuel cell system*

Possible architectures for a PEMFC system and a SOFC system are detailed here. Specific auxiliaries used for these systems will also be addressed.

2.3.2.1. *PEMFC*

Fuel supply circuit:

The anode compartment of a PEMFC is generally supplied with hydrogen from a storage tank. Hydrogen resulting from hydrocarbon (natural gas, methane, etc.) reforming can also be used. Nevertheless, the use of such architecture requires the addition of a reformer to this circuit. Moreover, non-negligible quantities of carbon monoxide and sulfur are present in the reformate at FC input. Or, these two gases can contaminate the membrane of a PEMFC. Hence, pure hydrogen is generally considered in PEMFC applications.

There are three operating modes for the hydrogen supply of the anode (Figure 2.5):

1) open anode mode;

2) closed anode or dead-end mode;

3) gas recirculation mode.

In the first case, a control system imposes the desired flow rate under a pressure that is substantially equal to cathode pressure, in order to limit the transverse mechanical stresses on the electrolyte membrane and consequently avoid its tearing. When this mode is used, the anode is oversupplied with hydrogen in order to properly cover the anode diffusion layers. It is then important to define the stoichiometric factor, FS_A (equation [2.12]), which corresponds to the ratio of the quantity of hydrogen sent to the anode and the quantity effectively required for the reaction. This configuration enables excellent operation of the FC, but diminishes the overall energy efficiency of the system, since hydrogen that is not consumed during the reaction is rejected in the atmosphere [PÉR 13]. This mode is generally used for the test bench characterization of FC or for stationary applications, as its implementation requires many sensors and actuators.

$$\dot{\eta}_{H_2} = FS_A \frac{NI_{PAC}}{2F} \qquad [2.12]$$

where F is the Faraday constant (F = 96,485 C/mol).

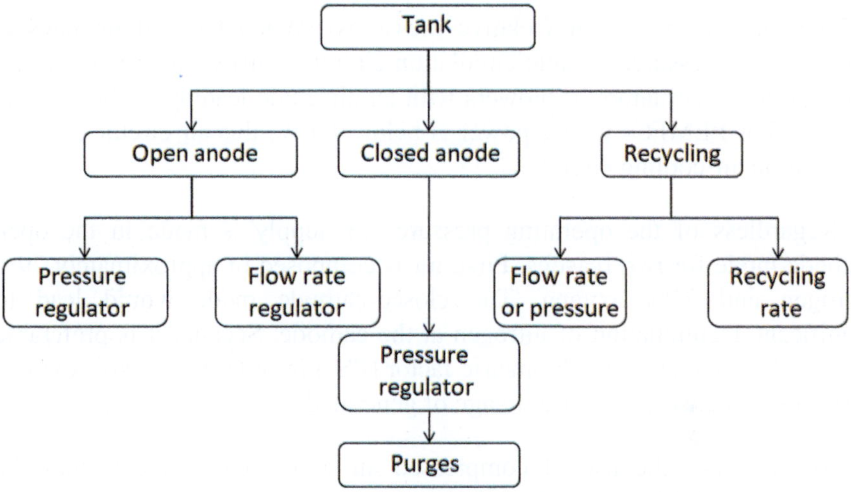

Figure 2.5. *Architecture of a fuel cell system*

In transport applications, the closed anode mode is more often used. Indeed, this architecture is the simplest, as it only requires a pressure regulator at the input and a discharge valve at the output. Regulating the input pressure, the FC only uses the hydrogen needed for the reaction depending on the required current. Actually, the stoichiometric coefficient is here equal to 1. Output purge is needed to avoid the accumulation of nitrogen and water resulting during the reaction. Thanks to the evacuation of these by-products, hydrogen is again present in a quantity that is sufficient to ensure the proper operation of the FC.

Finally, recirculation mode enables operation with a stoichiometric factor above 1, as the hydrogen that is not consumed at the output is now returned to the input by a pump or a recirculation compressor. This architecture requires the additional (with respect to the open anode mode) implementation of a specific recirculation compressor. In other words, as noted in the next section, using FC-dedicated compressors is not a common practice.

Oxidizer supply circuit:

Cathode side is supplied with oxygen in the air under a pressure that varies, depending on the manufacturers, from several hundred millibars to

1.5 bar and even 2.5 bar (relative pressures). When the cell operates at atmospheric pressure, a single circulation circuit is sufficient. It is generally composed of ventilators or blowers with an air guide leading to the cathode surface. For PEMFCs whose power is below 1 kW, this air circuit can also serve as an air cooling circuit.

Regardless of the operating pressure, air supply is made in the open cathode mode for two reasons. First, air is composed of approximately 79% nitrogen and 21% oxygen. The closed-cathode mode would lead to significant accumulation of nitrogen at the cathode. Second, it is preferable to use a high cathode stoichiometric factor (FS_C) in order to improve oxygen diffusion and ensure proper drainage of generated water.

Nevertheless, the use of compressed air at the cathode enhances the performance of FCs. For FCs whose power is above 1 kW, operation in pressurized mode is generally recommended. Compressed air is then obtained by means of a compressor unit. The latter supplies a variable air flow rate at various pressure levels (typically from 1.5 to 4 bars) [TEK 04]. It is possible to theoretically and physically prove that the efficiency of the electrochemical conversion will be all the higher as the gas partial pressures are significant. Figure 2.6 shows that an increase in pressure leads to an increase in the FC potential, which results in higher power density. Similarly, the Nernst equation (equation [2.13]) shows that the voltage is proportional to the logarithmic increase in pressure.

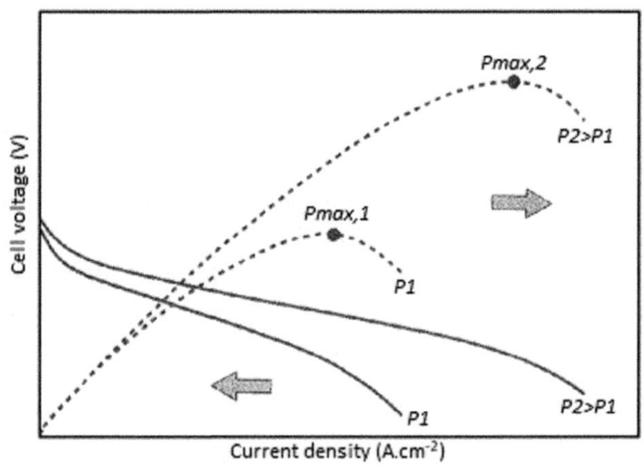

Figure 2.6. *Influence of air input pressure on voltage in a PEMFC*

$$\Delta V = \left(\frac{RT}{2F}\right)\ln\left(\frac{P_2}{P_1}\right) \quad [2.13]$$

where ΔV is the voltage gain of the FC (V), P_1 and P_2 are partial air pressures ($bar_{abs.}$), T is the reaction temperature (K), F is the Faraday constant (96,485 C/mol.) and R is the ideal gas constant (8,314.32 J/K/kmol.).

In the case of a high-pressure cell, a compressor unit is therefore necessary. For transport applications, it is essential to use compact and light air compressors that provide air without oil. However, they remain expensive, noisy and bulky, as besides the compression head, the engine, the control unit and the cooling system must be integrated.

An ANR (French National Research Agency) project has been undertaken by our team in order to investigate and characterize the various solutions available for the PEMFC systems with power above 10 kW. This project, named ICARE-CSP [ANR 10] (investigations and characterization and development of compressor for FC systems with powers above 10 kW), was conducted from 2008 to 2010.

These two studies have shown that it was very difficult for a compressor to meet all the following criteria (non-exhaustive list):

– required air quality, namely complete absence of oil, particularly for powers above 10 kW;

– acoustic discretion in a transport application, which is under strong constraints and subject to public acceptance;

– flow rate–pressure–electric power consumption compromise;

– high dynamics;

– overall dimensions and tolerance to mechanical loading;

– electromagnetic compatibility.

It was possible for us to propose in this context an exhaustive state of the art of the technological solutions for air compression for the PEMFCs on the market in 2010, to define generic specifications for air compression systems, to experimentally test and characterize the solutions on the market and to define the optimized control laws for air compressor units [DAV 10]. In particular, it has been possible to show three compressor technologies adapted to FC applications: double-screw compressor, centrifugal compressor and scroll compressor. However, none of these three compressors was fully compliant with the specifications imposed by the end user. Indeed, the double-screw compressor shows very high performances, but its acoustic signature is not adapted to transport applications (>120 dB at 3 m at rating). The centrifugal compressor cannot be used at low-power operating points due to its surge line, which renders its use very restrictive. Finally, the scroll compressor meets the aeraulics and energy criteria, but it is too bulky and heavy (> 30 kg for a 20 kW FC). Table 2.3 summarizes the main characteristics of the tested compressors [JEM 11].

Compressor type	***Max. flow rate. (g/s)***	***Efficiency***	***Max. pressure ($bar_{abs.}$)***	***Mass (kg)***	***Volume (dm^3)***	***Noise (dBA)***	***FC power (kW)***
Scroll	30	0.82	2.3	30	27	65	20
Double-screw	33	>1 with turbine	3.5	13	20	>120	20
Centrifugal	50	0.76	2.8	2.9	3.5	65	20

Green	Meets specifications
Yellow	Partially meets specifications
Red	Does not meet specifications

Table 2.3. *Comparative summary of various tested compressor technologies. For a color version of this table, see www.iste.co.uk/jemei/cells.zip*

With a double-screw compressor, efficiency (equation [2.14]) above 1 results from coupling a turbine for the recovery of hot air from FC output to the compressor unit shaft. This enables a gain ranging between +15% and 22% in terms of electric consumption:

$$\eta = \eta_v . \eta_{adiabatic,e} . \eta_{inverter} \qquad [2.14]$$

where η is the overall efficiency, η_v is the volume efficiency, $\eta_{adiabatic,e}$ is the experimental adiabatic efficiency and $\eta_{inverter}$ is the inverter efficiency.

The use of the compressor unit in the air circuit allows various supply control modes (Figure 2.7).

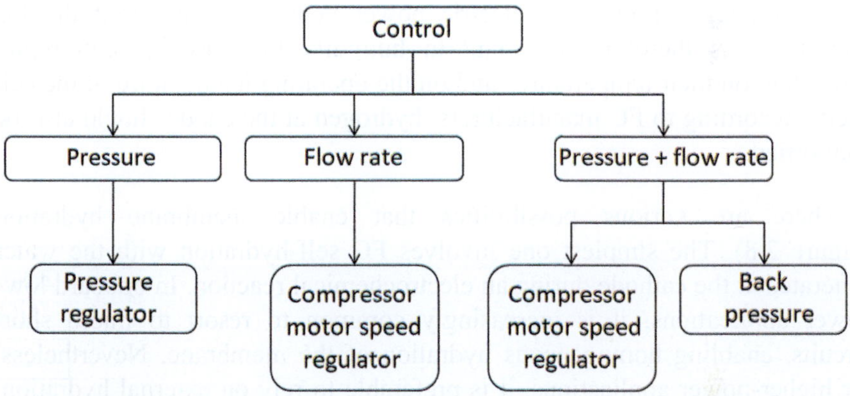

Figure 2.7. *Various control modes for the air circuit of a PEMFC*

The simplest mode is air flow rate regulation by acting on the speed of the engine driving the compressor. Nevertheless, for higher performances, it is preferable to regulate pressure and flow rate. Apart from the action on the compressor for flow rate, a backpressure valve should be placed at the FC cathode output.

The results obtained in the ICARE-CSP project show that the air supply circuit plays an essential role in the performance of the FC system. While the FC can respond with an excellent dynamic to the current demands imposed by the load, this is unfortunately not applicable to the compressor. Response times above 4 s have been obtained for passing from 0 to 90% of the desired air flow rate for a 20 kW FC. Thus, today, the air compressor remains one of the major technological obstacles of the FC system. Moreover, this circuit can get even more complex if coupled to air humidification. Indeed, air must be humidified to ensure proper proton conduction of the membrane. This process is generally conducted using a humidifier.

Humidification circuit:

In PEMFCs, water content of the membrane must be constantly controlled, in order to ensure the ionic conductivity of the membrane and facilitate proton (H^+) transport from the anode to the cathode. However, water excess can flood the active area of the electrode and hinder gas diffusion. This flooding and water plug phenomenon is particularly sensitive at the cathode, where water generation takes place. Dry-out and/or flooding of membranes diminish the electric power generation and limit the FC lifetime. It is therefore important to humidify the gases at cell input, depending on their temperatures and on the operating temperature of the cell itself. According to FC manufacturers, hydrogen at the anode should also be humidified.

There are various possibilities that enable membrane hydration (Figure 2.8). The simplest one involves FC self-hydration with the water generated at the cathode during an electrochemical reaction. In onboard low-power applications, it is increasingly common to resort to micro short circuits, enabling homogeneous hydration of the membrane. Nevertheless, for higher-power applications, it is preferable to rely on external hydration. Indeed, very high air flow rates at the cathode would pose the risk of drying out the membrane and diminishing FC performances.

Hence, various external solutions include enthalpy wheels for onboard generators, humidification systems based on water injection and/or bubbling for stationary systems and test benches. The overall objective for an onboard generator is the recovery of heat and water generated at the cell output in view of the upstream humidification of gases and thus reaching water balance, enabling autonomy at the system level [CAN 07].

For the best management of water content in the membrane and especially in order to avoid condensation phenomena, heat management of this circuit is paramount. It is often linked to the cooling circuit of the fuel cell stack.

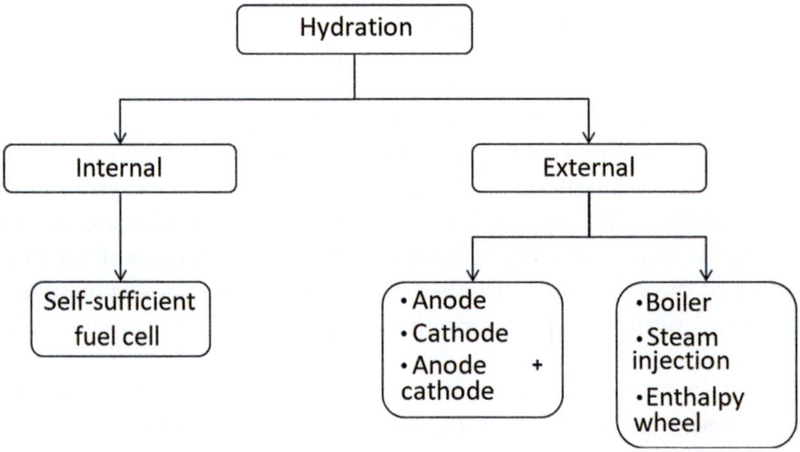

Figure 2.8. *Various modes for PEMFC hydration [CAN 13]*

Cooling circuit:

In addition to electric power, a PEMFC also generates heat. The cooling circuit evacuates the calories generated during the electrochemical reaction in order to control the operating temperature of the fuel cell (about 80°C). Temperature control of the PEMFC is essential, as system performance depends strongly on its temperature. Indeed, ionic resistivity of a dried-out membrane increases, and the voltage decreases across the FC. On the contrary, negative temperature could block reactions. In this case, the circuit becomes a preheating circuit and will need to be supplied by an external source.

Various cooling systems depend on the fuel cell power and the desired compactness. Various cooling topologies involve:

– natural convection with ambient air;

– forced convection using ventilators (bipolar plates can involve fins);

– circulation of a heat transfer fluid in the bipolar plates coupled with a heat exchanger (water/water or air/water). The fluid can be deionized water or glycol water in the case of a cold start or negative temperature.

Energy consumption of a FC system:

As already described in the previous sections, the PEMFC behavior is strongly influenced by the auxiliaries, ensuring its proper operation. Auxiliary consumption drastically reduces the overall energy efficiency of the system. These auxiliaries consume about 25–35% of the gross electric power available at FC output. Figure 2.9 shows the distribution of power consumed by various auxiliaries. Under optimal conditions, with an electric efficiency of the FC system of 65% (35% consumed by various auxiliaries), the efficiency of the FC generator unit drops to 32%.

Air supply circuit has the highest energy consumption. It draws around 15% of the power available at FC output. This is particularly due to the electric consumption of the engine driving the compression head of the compressor unit.

The other circuits, generally composed of pumps and valves, consume approximately 10% of the gross power. Although this section has shown that an optimized architecture may help reduce the consumption of auxiliaries, it is important to continue the research efforts especially on air compressors. The addition of an expansion turbine at the output of the cathode compartment coupled with the compressor shaft may help gain several points in efficiency. It may also prove interesting to couple humidification with the compressor by directly injecting water into the compression head. This will facilitate cooling and humidification of the compressed air.

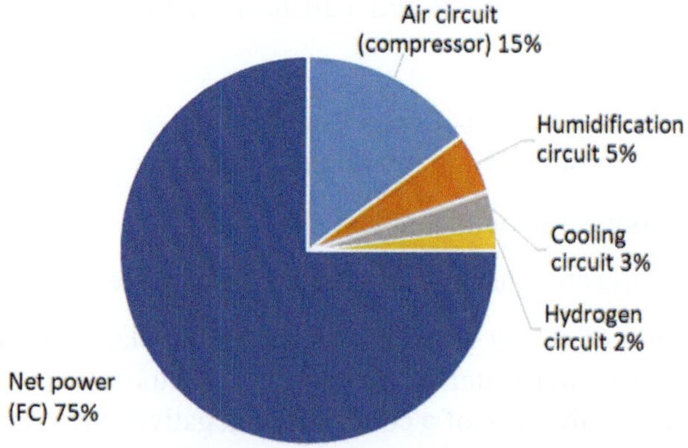

Figure 2.9. *Distribution of power consumed by the auxiliaries in a fuel cell system. For a color version of this figure, see www.iste.co.uk/jemei/cells.zip*

2.3.2.2. *SOFC*

High-temperature (> 700°C) operation of SOFCs involves relatively complex heat management of the system and control methodology. The "high-temperature" part, essentially composed of the thermal enclosure, involves many difficulties, particularly significant mechanical stresses on the sealing gaskets and SOFCs during thermal cycling. This may lead to reactant leakages or mechanical breakages of the fuel cells. The start times of SOFC systems are also a limiting factor that prevents them from being used in dynamic applications or in applications requiring frequent stop/start operations. Nevertheless, this technology offers interesting advantages for stationary use or combined heat and power generation.

Fuel supply circuit:

As previously noted, the anode of a SOFC can be supplied with pure hydrogen as well as with the gas resulted from hydrocarbon reforming (see Chapter 1, section 1.4.3.2). If pure hydrogen is used, then the circuit remains relatively simple, as it will send the gas to the anode under similar conditions to a PEMFC, with only the temperature varying. However, in SOFC applications, a reformer is most commonly used, as high-temperature operation enables the oxidation of CO resulting from reforming into CO_2. In this case, the supply circuit is essentially composed of this reformer. Several reforming techniques are possible, which have been introduced in Chapter 1.

Anode composition (Ni-YSZ cermet) in a SOFC requires specific use precautions, especially during the following operation phases:

– system start with stack heating up to its operating temperature;

– system stop with stack cooling;

– emergency stop of the system.

During these operating phases, and as long as the stack temperature remains above 400°C, the presence of air at the anode must be avoided in order to prevent the re-oxidation of the nickel contained in the anode layer. Sending hydrogenated argon to the anode during these phases contributes to avoiding this potential problem.

The anode circuit can only operate in open mode. As the stoichiometric coefficient is quite high, the exhaust gas is very rich in fuel at the FC output, which can be reused in various ways:

– in a burner, for heating one part of the system (reformer or air input at the cathode);

– it can be burned and the resulting heat can be used by means of an exchanger for combined heat and power generation. Energy recovery is all the more efficient as temperature is high;

– it can be used in a heat engine connected to a SOFC system (micro-turbine, Stirling engine, etc.) [GAY 10, PÉR 13].

It is worth noting that the gas entering the SOFC anode can be humidified according to the manufacturer recommendations.

Oxidizer supply circuit:

The oxidizer supply circuit brings the air to the cathode, similarly to a PEMFC system. An air compressor is generally used for air supply, as well as for controlling the temperature of the enclosure in which SOFC is confined. Nevertheless, for low-power applications (with small losses), a heat oven is generally necessary to preserve the operating temperature of the set.

Thermal circuit:

As noted previously, there are two possible thermal circuits. For high-power fuel cells, a reformer is generally used to produce the fuel. In this case, the stack and the reformer must reach a temperature of approximately 800°C in order to operate properly. Reformer and stack are then placed in a thermally insulated chamber. The increase in temperature is possible due to partial consumption (in a burner) of the fuel available at reformer input. Thermal cycles (passage from operating temperature to ambient temperature and vice versa) should be avoided in these systems, and it should be possible to maintain it at reduced temperature (about 400°C), thus avoiding premature ageing of the materials. This procedure obviously requires rigorous heat management and leads to additional (fuel or electric) consumption.

For lower-power fuel cells, an oven enabling the increase in system temperature should be used. Inside the oven, anode and cathode gases are brought to the stack (generally through pipe coils). These loops are added in the oven in order to heat the gases at high temperature before entering the stack. Figure 2.10 shows this architecture.

In this section, we have described the architectures of two most commonly used fuel cell system technologies. Regardless of the technology used and the system architecture, they must be experimentally implemented. The following section is dedicated to the presentation of various test benches implemented in our laboratory.

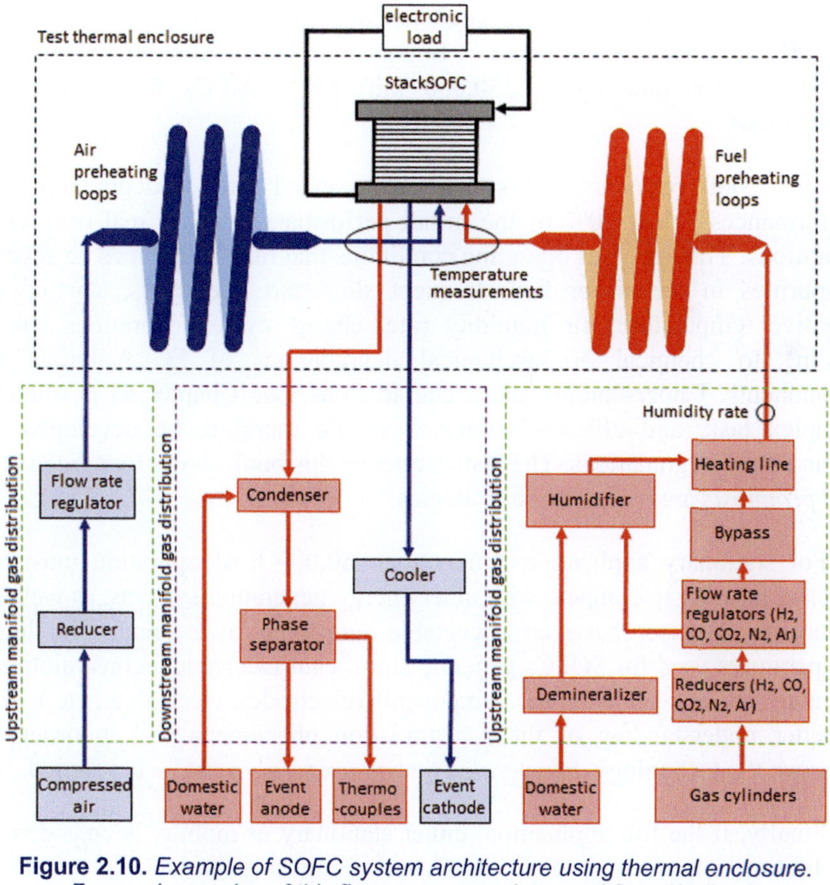

Figure 2.10. *Example of SOFC system architecture using thermal enclosure. For a color version of this figure, see www.iste.co.uk/jemei/cells.zip*

2.4. Limits of fuel cell systems

In the previous sections, we have presented fuel cell technologies that are expected to emerge very soon in transport or stationary applications. While the technologies seem mature enough for the market, a number of obstacles still need to be overcome. These are the limitations of the current FC systems, which are listed here.

The lifetime of PEMFCs is today approximately 3,000 h for a transport application. Considering an automobile application with an average speed of 50 km/h, this lifetime covers 150,000 km. The FC technology will be accepted by the consumers, provided that "FC vehicles" have the same and even higher performances compared to conventional vehicles. Hence, the DOE (United States Department of Energy) has set the lifetime objective at 5,000 h (corresponding to 250,000 km) for PEMFCs for automobile applications.

The 5,000 operating hours must be reached with a degradation of performances below 10% of the initial performances under real operating conditions. This involves operating conditions that may sometimes be severe (impurities in the air or fuel, frequent stop–start operations, start-up at negative temperature, air humidity rate, charge cycles, vibrations, etc.), leading to chemical or mechanical degradations of FC materials or components. Understanding these degradations (see Chapter 4) is often a complex task, and efficient strategies should therefore be developed to reduce these degradations. The last chapter of this book shows that *diagnosis and prognosis* are part of these strategies.

For stationary applications, more than 60,000 h of operation must be reached in order to compete with other energy generation systems, as well as to enable the user have an acceptable return on investment. The high temperatures used for SOFCs generate significant mechanical constraints on the materials as well as on the components (electrodes, electrolyte, etc.). For a better understanding of these degradation phenomena and in order to improve the technology, *accelerated testing protocols* must be developed.

Finally, if the full application, either stationary or mobile, is considered, the lifetime improvement of the auxiliaries composing the fuel cell system is also required.

The cost of fuel cell systems must be reduced, if they are to succeed in penetrating the market. There must be a reduction in materials and in the manufacturing costs of the fuel cell core components. The efforts should focus on developing high-performance membranes, low-cost catalysts requiring less precious metals and bipolar plates that are more resistant to corrosion, so that fuel cells become competitive.

It seems that for automobile applications, catalyst is the most expensive component of a FC with regard to mass production [JAM 10]. One solution would be the use of catalysts that do not require metals belonging to PGMs (platinum-group metals), which would reduce MEA costs. This solution is available for high-temperature fuel cells (SOFC and PAFC). Nevertheless, for these technologies, the efforts should concentrate on the cost of manufacturing cells, interconnectors or auxiliaries that should resist at high temperatures.

For an automobile application, the auxiliaries of the FC system represent approximately 50% of the cost of the full FC system (auxiliaries + FC). This proportion should also be reduced. Nowadays, certain components are specifically designed for FC technology and power. Possible examples include the air compressor, exchangers, humidifiers and even the static converter.

As for the air compressor, we are at present in an advanced stage of discussions with the designers of electric machines and with the manufacturers who are developing these machines and the compression heads. Joint design, in cooperation with experts in electrical engineering, on the one hand, and with experts in the fluid mechanics of rotating machines, on the other hand, is important, as a very high-speed (> 100,000 rpm) compressor can be used to obtain a miniaturized object with low energy consumption.

In order to penetrate the market, *the efficiency and performances* of the fuel cell system must be above or equal to those obtained with competing technologies. Fuel cell performances first influence the efficiency of the generator. Nevertheless, system auxiliaries are also responsible for performance losses.

The management of water generated for low-temperature FCs must enable optimal operation of the FC over a wide range of operating temperatures. Flooding or dry-out of the membranes should particularly be avoided, as it may block the electrochemical reaction (flooding) or reduce the proton conductivity of the membrane (dry out). Moreover, for stationary or transport applications, the FC system must be able to operate at negative ambient temperatures. Thus, achieving better management of water and operating at negative temperatures require an improvement in the design of gas diffusion layers, channels for the distribution of bipolar plate gases, catalytic layers and membranes.

FC cooling and gas humidification are also key points in achieving better performances. Moreover, less voluminous and more efficient humidification systems must be developed. For PEMFCs, the low gradient between the operating temperature and the ambient temperature does not permit the optimal use of the heat generated during the reaction, which renders the cooling systems very voluminous, as they require implementation of significant heat exchangers. On the contrary, for SOFCs, the use of the generated thermal energy improves the overall system efficiency, but may lead to design complications of the system for the extraction of heat generated at very high temperatures. Finally, it is essential to have efficient stop–start protocols for water management at negative temperature.

Air supply ensured by compressor must be optimized. As noted in this chapter, the compressor can use up to 15% of the power supplied by the FC. Improvement of this actuator should consider its compactness, efficiency and cost.

Finally, the profile of the electric load applied to the FC is one of the parameters with the strongest influence on system performances. A transport application should start very rapidly, irrespective of weather conditions, or respond to strong power dynamics. For stationary applications, and particularly for SOFCs, rapid starts, sudden stops and thermal cycling should be anticipated to avoid damaging of materials. This involves a reduction in dynamics and over-dimensioning of stacks to accommodate these thermal constraints.

To overcome these problems, hybridization of generators seems to be one of the promising solutions. This will be the focus of Chapter 3.

2.5. Conclusion

In this chapter, we have provided elements that are essential for understanding the operation of a FC and its system. After having described the various technologies of the existing FCs, we have focused on two most commonly used FCs nowadays, namely the PEMFC and the SOFC. PEMFCs can be used for transport and/or stationary applications, while SOFCs seem dedicated to the stationary market, although in this chapter, we have shown that transport applications can also be considered. Regardless of the technology used and the retained application, the system aspect deserves attention. We have highlighted the strong interactions between fuel cell and its auxiliaries, rendering the system complex, multi-physical and multi-scale. Nevertheless, fuel cell generators have many advantages, particularly in terms of efficiency, acoustic signature and polluting emissions.

The various studies and experimental implementations developed in this chapter have enabled us to highlight the subject of FC system dynamics. While the FC is very dynamic, its auxiliaries (air compressor, H_2 supply, humidification, cooling, etc.) have significantly different response times (from several milliseconds to several minutes), putting the overall operation at a disadvantage.

As will be noted in Chapter 3, hybridization of FC generators with other energy sources seems an essential alternative. Besides improving the performances of the FC system, hybridization also enables a significant increase in lifetime.

3

Hybridization of Generators

3.1. Introduction

The improvement of fuel cell generator performance remains one of our major preoccupations, and the use of such technology requires hybridization. The immediate focus is on transport applications, in which dynamic constraints can be sources of degradation of the fuel cell stack. The use of the fuel cell as single energy source in an electric vehicle requires precise and time-consuming control command of the fuel cell system, particularly when there is no *a priori* knowledge of the mission profile to be followed. Even though the fuel cell has very short electrochemical response times, the surrounding auxiliaries limit the overall dynamics of the system. The air supply of the cathode is the best illustration. The dynamics of the air compressor group is generally slower than the intrinsic dynamics of the fuel cell. Performances are consequently reduced.

Therefore, in order to reduce the overall energy consumption of the electric vehicle, it is important to be able to (even partially) recover the braking energy, which is impossible only with the fuel cell, as the generator is not reversible. This brings us to the limits of the fuel cell [HIS 06].

In fact, hybridization of this FC with energy storage devices must be considered in order to reach high overall efficiencies.

This research path, which is out of necessity more systemic, was the subject of the doctoral theses by Mona Ibrahim [IBR 13] and Jérôme Baert [BAE 13a].

3.2. Hybridization of electric power sources

In embedded or stationary systems, there is an evolution towards hybridization of multiple sources of various natures. This hybrid solution is particularly interesting, since it enables various complementary combinations in terms of energy and power of the various sources implemented, which leads to certain advantages in terms of performances. These hybrid systems are designed for the generation and use of electricity. They are also often used for energy storage, which in this case are referred to as electric energy storage systems (EESS).

In stationary applications, hybrid systems are generally independent of the electric network and include more than one energy or power source. Their power can vary from several megawatts (for an island) to several kilowatts (for a chalet or a small isolated house). Hybrid systems often rely on renewable energies, which are, by definition, intermittent. Hence, it is important to have high-performance electric storage systems. Hybridization of these various sources and/or storages offers various advantages (efficiency, lifetime, energy density and power, etc.), in terms of performance, stemming from each individual element.

For onboard applications, hybrid systems are mainly used in transport, particularly in electric and hybrid vehicles. This chapter focuses on this type of application.

3.2.1. *Hybridization for transport-oriented systems*

The American expression "*Two heads are better than one*" is readily applicable to vehicle drivetrain. Using a combination of internal combustion engine (ICE) and electric machine powered by energy stored elsewhere, hybrid vehicles have reached very high performances on the market in terms of reduced polluting emissions.

It is worth mentioning the performance criteria commonly used for the classification of hybrid systems. According to [ALL 10, BEN 05], there are six conventional criteria:

1) specific power (W/kg) is the power that an element weighing 1 kg can supply;

2) specific energy (Wh/kg) is the energy that an element weighing 1 kg can supply. It is worth noting that specific energy has a maximum theoretical value that can never be reached. This is essentially due to the fact that, in general, the storage elements are not fully dischargeable;

3) durability is characterized by the number of charge/discharge cycles that the electrochemical storage elements (battery) or electrostatic storage elements (ultracapacitor) can handle. For systems with external energy storage (electrolyzer and/or FC), it is defined by the number of "normal" operating hours (as long as performances remain below a threshold to be set);

4) the cost (€/kWh) is the initial investment price plus the cost involved by the use of the element;

5) recyclability should be taken into account at the end of the element's life;

6) security reflects the risk(s) associated with a storage element or with an electric power source.

The energy sources in a hybrid vehicle can have various natures (electrochemical, mechanical, thermal, etc.) and characteristics. The following are the components of a hybrid vehicle:

– internal combustion engine (ICE);

– electrochemical accumulator (EA);

– ultracapacitor (UC);

– fuel cell (FC);

– inertia wheel.

These sources are chosen depending on the specifications of the intended application, and particularly on the energy density and power density of the element. The Ragone diagram (Figure 3.1) enables a performance-based comparison among energy storage elements. According to this diagram, in order to cover a wide range of energy density and power density, hybridization of various technologies is obviously needed.

The choice of elements to hybridize aboard a vehicle is essential. The hybrid electric vehicle (HEV) is minimally composed of an internal combustion engine and an electric machine, the aim being to reach the performances of a conventional vehicle (ICE) and of an all-electric vehicle (EV). Compared to the conventional vehicle, the HEV must reduce local polluting emissions as well as provide better autonomy than the EV.

It is worth noting the existence of various HEV architectures. They can be classified into the following three categories:

– Series hybrid (Figure 3.2(a)): traction power stems entirely from electric energy. The primary source, which can be an ICE (or a FC), is not mechanically coupled with the drive shaft, but with an electric generator. This architecture presents a real advantage when used with an ICE. Indeed, the latter is never used in idle mode, which reduces the polluting emissions, but can be used at its rated operation point (reduction of emissions). Nevertheless, this architecture requires a large number of onboard batteries, which involves significant mass and volume.

– Parallel hybrid (Figure 3.2(b)): the interest of this configuration lies in its classic traction chain (ICE + transmission) and an electric traction chain in parallel (batteries + generator coupled to wheels). Each of these technologies can independently or simultaneously operate to increase the supplied power.

– Series–parallel hybrid (Figure 3.2(c)): as the name indicates, this architecture is a combination of series and parallel configurations. The ICE directly drives the vehicle or provides electric power to the batteries through the electric generator.

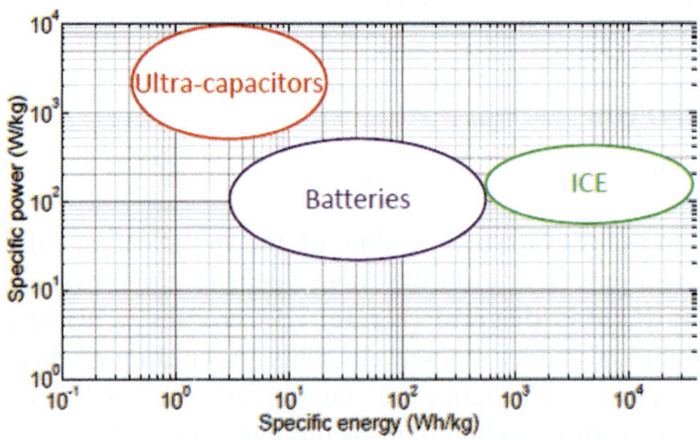

Figure 3.1. *Ragone diagram [BAE 13a]. For a color version of this figure, see www.iste.co.uk/jemei/cells.zip*

The retained architecture varies depending on the type of vehicle (light-duty vehicle, truck, locomotive, helicopter, etc.) to be designed. This choice involves also a control structure and adapted energy management. This will be the focus of the next section.

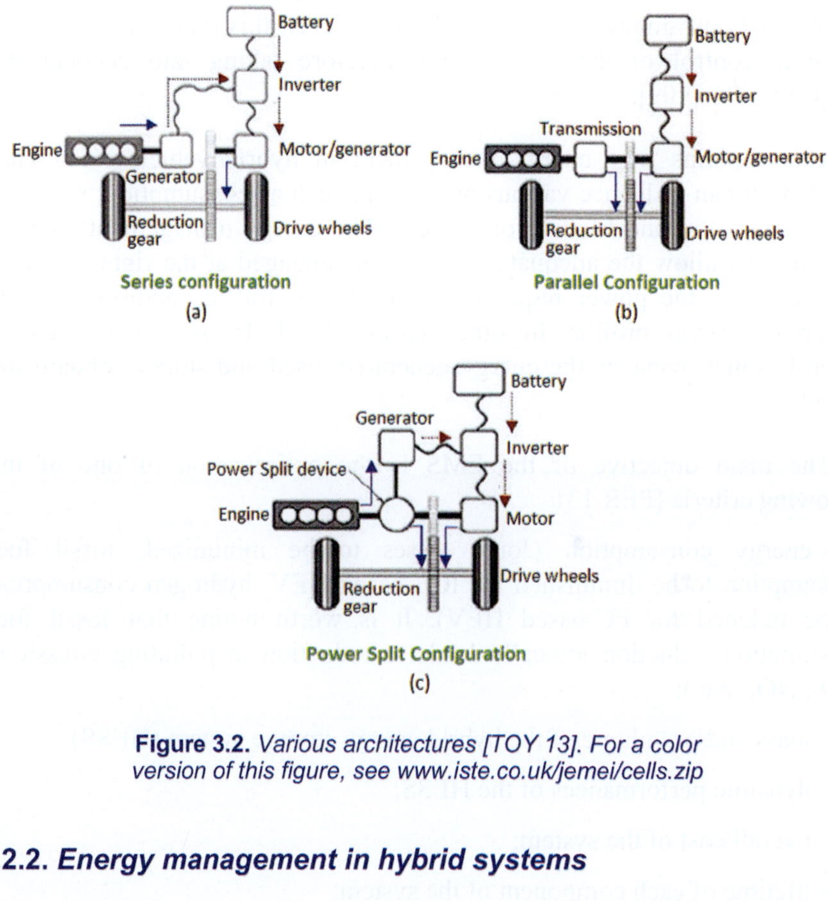

Figure 3.2. *Various architectures [TOY 13]. For a color version of this figure, see www.iste.co.uk/jemei/cells.zip*

3.2.2. *Energy management in hybrid systems*

The presence of various energy sources in a vehicle involves the management of the local energy of each element, enabling performance improvement. From a systemic point of view, control of power flows among various sources is required in order to increase the overall efficiency of the system and to ensure proper operation. This is referred to as the energy management strategy.

Local energy management amounts to optimizing the operation of a given element or subsystem, while overall management boils down to optimizing the overall system operation, including all underlying contradictions. While overall energy management most often takes the form of control strategy studies based on quasi-static models, it also implies rigorous consideration of the interactions among subsystems. At local level, this relates to aspects of dynamic control of the system and therefore taking into account the transients [BOU 09].

This explains why energy management in hybrid vehicles is a vital problem. It can influence various parameters such as consumption, sizing of energy source(s) and lifetime of sources. The energy management strategy (EMS) must allow the adequate source to be engaged at the right moment, depending on the power required by the driver, traffic conditions or the required mission profile. In other terms, the EMS is a mathematical algorithm that manages the energy generated, used and stored onboard the vehicle.

The main objective of the EMS is the optimization of one of the following criteria [PÉR 13]:

– energy consumption (Joule losses to be minimized, fossil fuel consumption to be diminished for ICE-based HEV, hydrogen consumption to be reduced for FC-based HEV). It is worth noting that fossil fuel consumption reduction logically leads to reduction in polluting emissions (CO_2, NO_x, etc.);

– mass and/or volume of the hybrid energy storage system (HESS);

– dynamic performances of the HESS;

– overall cost of the system;

– lifetime of each component of the system;

– security/safety.

Simultaneous optimization of several of these criteria is obviously possible. A function including the various weighted criteria needs to be defined.

There are two options for the optimization of management strategies:

Offline optimization can be achieved when the driving cycle is *a priori* known. In the case of VEHs application, these methods aim to minimize the energy consumption and the polluting emissions of these systems. Dynamic programming, Pontryagin's minimum principle, Lagrange multiplier and constrained optimization of Hamiltonian are highly efficient techniques [BAE 13a]. Their implementation requires working on the basis of a well-known mission profile. The resulting energy management strategy is therefore only optimal for the mission profile (and even for a vehicle architecture), which will never be the same under real conditions. The optimization processes used by these techniques are difficult to implement. The larger the number of constraints to be considered, the more complex the optimization becomes and the longer the computation time is [CAU 10, LES 10]. Hence, this type of strategy is not suitable for real-time applications and requires *a priori* knowledge of the mission profile.

These techniques are not presented in detail here, as they do not seem adapted to our research objectives.

Online optimization is applicable when the mission profile is not known in advance. These are real-time methods that propose suboptimal solutions. They rely on the optimization principle that takes into account the entire studied system and neglects certain constraints in order to lighten the optimization process. Unlike offline techniques, they do not require *a priori* knowledge of the mission profile. Examples of these methods include fuzzy logic, neural networks, predictive control and the wavelet transform.

An in-depth presentation of these methods is provided in the next section.

3.2.2.1. *Online strategies*

Strategies based on online optimization algorithms provide overall optimization, taking the full system into account. These methods are carried out in real-time. They only consider the real-time information on the system, such as speed, acceleration, power, voltages, currents or the states of charge of accumulators [HAJ 06]. These techniques enable online optimization without *a priori* knowledge of the mission profile. They may not yield a necessarily optimal final solution, but it is a realistic solution [KER 09].

Fuzzy logic:

In recent years, fuzzy logic has been implemented in a wide variety of applications. It has rapidly become one of the most used methods for the development of control/command laws for complex and sophisticated systems. The fuzzy logic approach facilitates the translation of human knowledge and behavior into computer language [JAM 09a]. It bridges the gap between purely mathematical approaches and purely logic-based approaches. While certain techniques use complex and precise equations in order to model real phenomena and behaviors, fuzzy logic satisfies both human logic and human behavior [LIA 08]. It is successfully used in several scientific fields such as control [JAM 09b], navigation [SCH 12] and energy management [BAE 13a, SOL 11]. The techniques that use type-1 and type-2 fuzzy logics are detailed in section 4.4.

Neural networks:

Neural networks (NN) are models that draw their inspiration from the biology of the human brain. They enable the replication or acquisition of human behavior and knowledge and making "intelligent" decisions. The NN are composed of several interconnected artificial neurons that process information and provide results.

An artificial neuron is a nonlinear parameterized algebraic function with bounded values. The neuron operates variables that are usually referred to as neuron inputs, and the value of the function is referred to as the output [DRE 02].

The neuron receives a combination of inputs (variables x_i) and weights them by means of real values called synaptic coefficients or synaptic weights (parameters w_i). If these coefficients are positive, synapses are excitatory, and if they are negative, synapses are inhibitory. The neuron thus computes a weighted sum (called potential ν) of its inputs equation [3.1]. This potential ν to which a constant term or bias denoted here by "b" is added becomes the argument of an activation function f that computes the output y equation [3.2]:

$$v = w_0.b + \sum_{i=1}^{S} w_i x_i \quad [3.1]$$

$$y = f(v) \quad [3.2]$$

The function f can be parameterized in various manners, depending on the use of the neuron and on the continuous, discrete or binary nature of the neuron state y. The most widely used activation functions are sigmoid and linear functions.

Therefore, a neuron simply realizes a linear or nonlinear parameterized function of its input variables. Neurons are important due to the resulting properties of their association in networks. There are two types of neural networks: non-recurrent neural networks and recurrent neural networks. These networks have one or more layers of neurons. In the latter case, they are called multilayer neural networks. The number of layers and of neurons per layer depends on the nonlinearity of the problem at hand. A learning phase enables the determination neuron's weight and the implicit acquisition of knowledge of the system at hand. Weight calculation aims to minimize the error between desired and calculated outputs. Thus, neural networks are likely to yield efficient and elegant solutions in various fields: shape recognition, modeling of industrial processes, static or dynamic modeling of nonlinear processes, environmental control, robotics, bioengineering, time series prediction, process control and so on [JEM 04, JEM 08b].

Predictive control:

There are three classes of competitive approaches in the field of predictive control. The classification relies on three criteria: optimality, control horizon and optimization objective. Most control strategies rely on heuristic controllers. Heuristic approaches require significant computation times and often yield suboptimal results [KUT 10]. In the context of predictive control, the best categories of optimization algorithms use dynamic programming techniques. These techniques require long computation time, which is not really suitable for real-time implementation in vehicle control strategies.

Wavelet transform:

The wavelet transform decomposes an original signal into components of various positions and scales [MAL 08]. It makes the simultaneous extraction of information in the time and frequency domains possible (in contrast to other transforms, such as the Fourier transform, which can only extract information in the frequency domain). This approach is notably used to identify frequencies in the real-time dynamic power profiles. Using this algorithm, power demands are distributed to available energy sources depending on their optimal frequency ranges.

Moreover, thanks to their localization properties, wavelet transforms are very useful in the analysis and detection of non-stationarities and variations in a given signal.

A detailed description of this method is given in section 4.3.2.2.

3.2.2.2. *Example of ECCE vehicle*

The ECCE vehicle or "electrical chain components evaluation vehicle" (Figure 3.3) is a heavy hybrid vehicle (14 tons), designed for real-time evaluation of electrical components such as electric machines, static converters, energy sources or energy management strategies. Its traction chain comprises four independently controlled electric machines. The ECCE vehicle is a mobile test bench that can feature various energy sources such as a fuel cell system, electrochemical accumulators, ultracapacitors, an inertia wheel and even an internal combustion engine.

Started in 1997, the ECCE is a research project financed by the Directorate General of Armaments. This project has involved substantial work aimed at building and evaluating the vehicle (evaluation of the electric machine in [ESP 06] and of the battery in [KAD 09] and the study of energy management of the traction chain in [PUS 02]). Since 2008, this project has focused on energy control and management in the presence of various sources of energy (fuel cell, ultracapacitor, inertia wheel, internal combustion engine, etc.).

In Javier Solano's thesis [SOL 12b], an energy flow management strategy was implemented on the vehicle. Experimental results collected (Figure 3.4) on the road show that the power required for mission completion is supplied by three sources: FC, ultracapacitors and batteries. Let us primarily note that bus voltage is stabilized at approximately 540 V. Several variations generated by the charge/discharge of batteries (directly connected on the continuous bus) are found. Ultracapacitors prove to be able to perfectly respond to the dynamics imposed by the mission profile and to recover the braking energy. The power supplied by the FC is adapted to the gas supply dynamics and notably to the air circuit comprising the compressor. Experimental validation on a hybrid vehicle of the energy management based on type-2 fuzzy logic constitutes, to date, a world premiere. In the following section, experimental data yielded by ECCE are used for developing a new energy management.

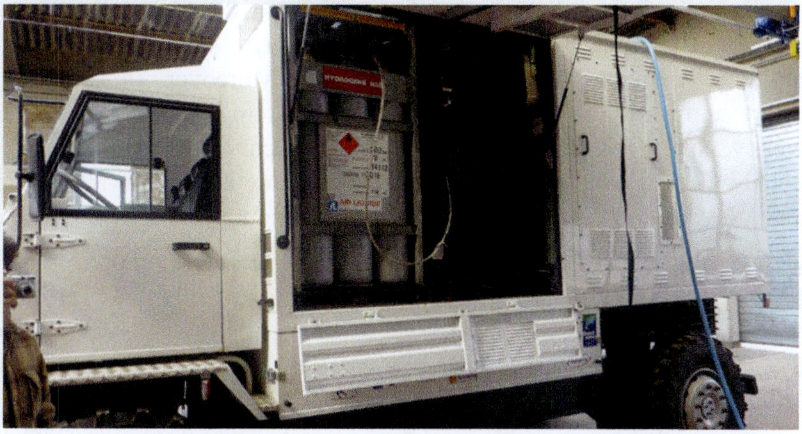

Figure 3.3. *Hybrid ECCE vehicle. For a color version of this figure, see www.iste.co.uk/jemei/cells.zip*

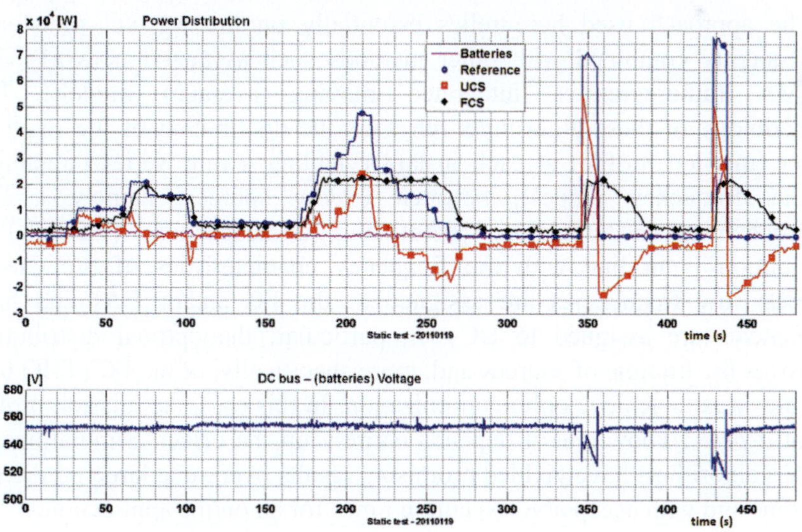

Figure 3.4. *Experimental results. For a color version of this figure, see www.iste.co.uk/jemei/cells.zip*

3.3. Hybridization of fuel cell generators

Hybridization of the fuel cell generator is essential for increasing its lifetime. In fact, the frequency range that permits secure FC operation is relatively limited (low frequencies from 10^{-4} to 10^{-2} Hz). It is therefore preferable to hybridize the FC with other energy sources, enabling the coverage of a much larger frequency range. Moreover, in the case of transport application, the use of a hybrid system improves the overall efficiency, since the braking energy can be recovered, which cannot be expected with only one FC, as it is a voltage source that is non-reversible in current. Nevertheless, the use of various energy sources in a hybrid vehicle, of various natures and characteristics (as in an ECCE vehicle, for example), requires a strategy for power distribution according to the characteristics of each source.

This is the subject of the thesis by Mona Ibrahim [IBR 13]. The use of a hybrid system requires energy management implementation. The objective of this thesis was therefore to realize a real-time energy management strategy for a hybrid vehicle. The experiments referred to in this thesis have been conducted on an ECCE vehicle.

The approach used here relies essentially on the wavelet transform, subsequently combined with other methods such as neural networks and ARIMA (autoregressive integrated moving average) models. The methodology of these algorithms relies on the distribution of the required power (mission profile) at each instant (yielding a time series containing various frequency ranges), between energy sources, namely a FC, electrochemical accumulators (EAs) and ultracapacitors (UCs). This distribution is in line with the frequency characteristics of the energy sources: low frequencies are assigned to the FC and/or EAs and high frequencies are assigned to UCs. In particular, the optimal distribution improves the lifetime of sources and, more specifically, of the FC [THO 09]. Our aim was to achieve online optimization. In this case, the mission profile is not *a priori* known, as it depends entirely on the driver. Simulations of various algorithms have yielded promising results over quite large prediction horizons and with acceptable execution times for an online application.

In the first part, the wavelet transform will be rapidly introduced. The second part will show how this mathematical method can be applied to energy management for hybrid vehicles. The last two parts show that power profile prediction is required in certain cases, in order to achieve high-performance online management. Neural networks and NARNN or ARIMA models make it possible to reach this objective. The set of results presented in these parts draw on the research work of Mona Ibrahim [IBR 13].

3.3.1. *Wavelet transform application to energy management*

3.3.1.1. *Wavelet transform presentation*

Time-scale methods such as wavelet analysis are tools adapted to the analysis of non-stationary signals. Similar to the Fourier transform, the wavelet transform enables passage from a time representation to a frequency representation, and similar to window transforms, it permits the measurement of the time variations of the frequency components of a signal. Nevertheless, time-frequency resolution of the wavelet transform is more flexible and better optimized [MAL 99]. This method relies on the ability to adapt the window (wavelet) to the scale of analysis: for low frequencies, fine time resolution is not required and therefore wide time windows and narrow frequency windows are used, whereas for high frequencies, wide frequency windows that are time-localized are used instead. This adaptive aspect of the window stems from the fact that only one function generates at each scale a family of “windows” by compression and dilation. In other terms, the narrower the window in time, the more compressed and rapidly oscillating the wavelet, and the wider the window, the more dilated the wavelet. The use of an adaptable family enables us to dispense with the compromise between time resolution and frequency resolution. The wavelets have drawn increasing interest over the last 20 years. After being confined to the field of mathematics for a long time, they have been adopted in signal processing as a result of Mallat [MAL 99] establishing the connection with filters. Today, they are a link between mathematicians and engineers. The wavelet transform has been applied in various and diverse fields, such as noise elimination, image compression and analysis of seismic signals and signals emerging from the medical field [CHE 06, SAI 95]. It is part of the family of methods that allow the exploration of the signal, including its characteristics at various scales, which seems well adapted to signals generated by energy sources. In the context of our study, the focus is on the mission profile of the vehicle power. This signal is composed of a rich collection of information, which is useful to separate in view of the power distribution among various sources of energy.

Several theoretical elements required for the implementation of wavelet decomposition are presented here.

A wavelet is a time-localized mathematical function ψ, which transforms the signal to be analyzed so that the information contained at various scales can be evidenced. It owes its name to its oscillating and compact character. Any mathematical function can be considered a wavelet provided that it is oscillating and has finite energy and zero average [MIS 07a]. Figure 3.5 shows several examples of wavelets.

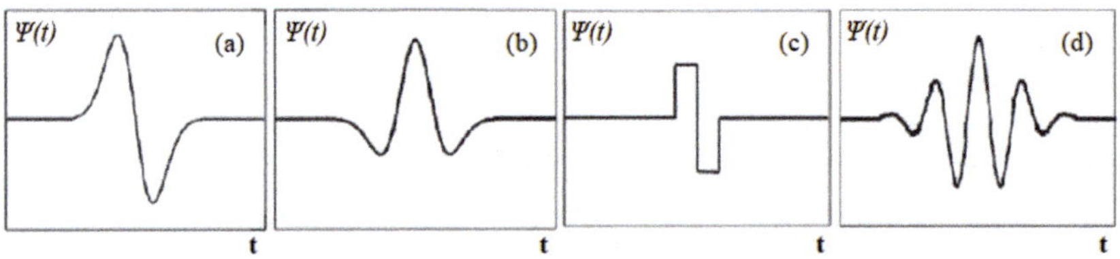

Figure 3.5. *Examples of wavelets: a) Gaussian, b) Mexican hat, c) Haar and d) Morlet [ADD 02]*

Based on this "mother wavelet", a wavelet family $(\psi_{a,b}(t))_{a,b}$ can be defined by time translation and dilation (or compression) according to relation 3.3:

$$\psi_{a,b} = \frac{1}{\sqrt{a}}\psi\left(\frac{t-b}{a}\right) \qquad [3.3]$$

where b, called the "time translation factor", describes the time axis and a, called the "scale variable", enables shifting in various scales.

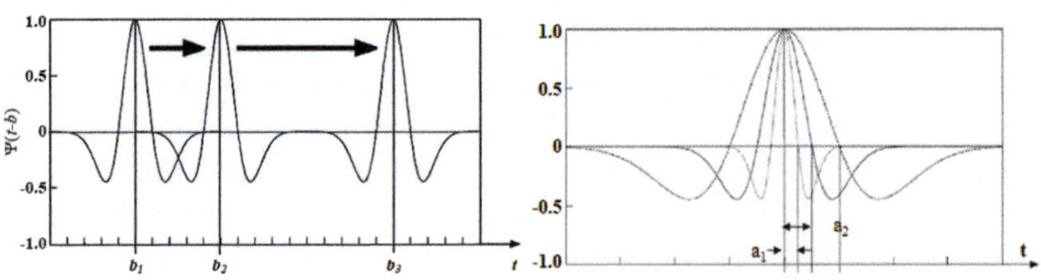

Figure 3.6. *Illustration of translation (a) and compression/dilation (b) notions*

Figure 3.6 shows the wavelet at various time locations (b_1, b_2,...), (a) and in various states of compression and dilation (a_1, a_2,...) (b).

For the best analysis of a signal, it is preferable to use the continuous wavelet transform (CWT). The CWT of a signal $f(t)$ is defined as the integral of the product of $f(t)$ (signal to be analyzed) by the family of analyzing functions. This transform generates a set of coefficients $C_f(a,b)$ called "wavelet coefficients":

$$C_f(a,b) = \int_R f(t)\psi_{a,b}^*(t)\,dt \qquad [3.4]$$

where * designates the conjugate. Compared to Fourier analysis, the analyzing function is no longer a "fixed" sinusoid, but a set of functions $(\psi_{a,b}(t))_{a,b}$ adaptable to various scales a and times b.

The signal $f(t)$ is then fully described by the coefficients $C_f(a,b)$. From a mathematical stance, conducting a CWT of a signal amounts to multiplying a signal by the family of basic functions and calculating the degree of correlation between this signal and the analyzing wavelet equation [3.4]. In practice, for each scale variable a, the wavelet is moved from the origin of time axis along the quantity to be analyzed (by varying the translation variable b), in order to calculate a series of correlations between the two. The wavelet takes various forms depending on the values of a and various positions depending on the value of b. The results of these correlations correspond to a set of wavelet coefficients, $C_f(a,b)$ at each scale and with each translation. These are all the larger as the wavelet shape approaches that of the signal to be analyzed.

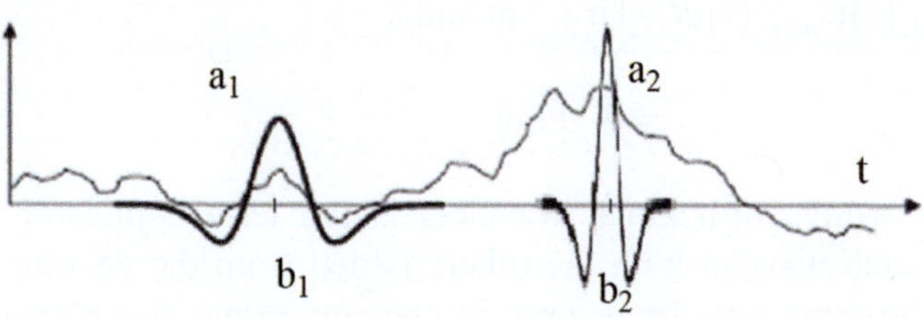

Figure 3.7. *Continuous transformation into wavelets*

Figure 3.7 illustrates this procedure for two scale factors, a_1 and a_2, and for two translations, b_1 and b_2. Parameter b indicates the time position of the "window", and the scale factor a gives the frequency resolution. The scale factor a can be interpreted as a road map scale: if a is small, then there is poor frequency resolution (therefore good time resolution), and if a is large, the opposite is true.

CWT provides redundant information that increases computation time due to the continuous variation of a and b. Discrete wavelets eliminate this redundancy by using discrete scale and translation values equation [3.5]. This is referred to as "decomposition into wavelet series":

$$a = a_0^{\,j} \qquad \text{with } a_0 > 1$$

$$b = k.b_0.a_0^{\,j} \qquad \text{with } b_0 > 0\text{, j and k integers.} \qquad [3.5]$$

In many applications [MIS 07a], the values of b_0 and a_0 are limited to 1 and 2, respectively. This decomposition is known as dyadic wavelet decomposition and its principle involves building wavelets ψ that generate orthonormal bases of L^2 (R) (space of square integrable or finite energy functions) by translation and dilation [MAL 09]:

$$\psi_{j,k}(t) = \frac{1}{\sqrt{2^j}}\psi\left(\frac{t - 2^j k}{2^j}\right), (j,k) \in Z^2 \qquad [3.6]$$

The notion of orthogonality is verified (only for discrete wavelets), if:

$$\int_R \psi_{j,k}(t)\psi_{m,n}^{*}(t)\,dt = 1 \text{ if j = m and k = n} \qquad [3.7]$$

$$= 0 \ \text{ if not}$$

In other words, orthogonal wavelets is the term applicable when the product of each wavelet with the others issued from the decomposition on basis 2 of the same wavelet is zero. In concrete terms, this means that they are independent and contain mutually exclusive information. Then, the decomposed signal $f(t)$ can be rebuilt as follows, with no information loss, relying on this orthogonal wavelet basis:

$$f(t)=\sum_{j\in Z}\sum_{k\in Z}C_{j,k}\psi_{j,k}(t) \qquad [3.8]$$

where $C_{j,k}$ are coefficients resulting from the "direct" decomposition into wavelets. This formula corresponds to the "inverse wavelet transform", in the case of the discrete transform. Discrete dyadic orthogonal wavelets are associated with a scale function ϕ defined similarly to the wavelet family by:

$$\phi_{j,k}(t)=\frac{1}{\sqrt{2^j}}\phi\left(\frac{t-2^j k}{2^j}\right),(j,k)\in Z^2 \text{ with } \int_{-\infty}^{+\infty}\phi_{0,0}(t)dt=1 \qquad [3.9]$$

This family is applied to the signal to yield "approximation coefficients" (a "smoothed" or "locally averaged" version of the signal). Let us note that the coefficients generated by the wavelet application equation [3.6] yield "detail" coefficients.

From a purely signal processing perspective, there are only two types of functions: the signal to be analyzed and the function that analyzes or filters it. Further on, the CWT of a signal can be seen as the passage of this signal through a "filter bench": at a given resolution, a low-pass filter (h) (associated with the scale function) gives a coarse image of the signal, and a high-pass filter (g) (associated with the wavelet) gives the details. These two filters are complementary: frequencies cut by one of them are recovered by the other. This proves that high-frequency and low-frequency notions are related to a given scale. Frequencies considered high at a given scale may be considered low at the next scale, that is, at a twice finer resolution. The technique based on filter benches is a "classical" practice in the field of signal processing, which S. Mallat relied on when proposing a link between wavelet decomposition and filters [MAL 09]. This combination served to implement a rapid orthogonal wavelet transform based on filters. Thanks to this different approach, the wavelet transform can be used easily. (equation [3.6]) [YOU 09c].

As an illustration of these theoretical explanations, the following part describes the results obtained.

3.3.1.2. *Application of the wavelet transform to energy management in a hybrid vehicle [IBR 13]*

The focus here is on the application of WT to ECCE vehicle. Various possible combinations of energy sources are taken into account. The specific endeavor is the lifetime increase for FCs, as well as for batteries. Indeed, similar to the FC, batteries are subject to degradations due to high current dynamics. Therefore, addition of ultracapacitors is recommended, as an aid to the fuel cell and to the batteries, in order to cover the high-frequency parts of the current (or power) demand.

WT is used here for decomposing the power profile of the vehicle into an approximation signal (low frequencies) and detail signals (high frequencies).

This requires WT parameterizing, which involves more specifically choosing the mother wavelet of the decomposition level.

Choice of mother wavelet:

The choice of the mother wavelet relies on many criteria, depending on the considered application. This choice directly influences the analysis of the transient signal. According to the literature [GAO 11, ISA 12, SAL 05, YAN 07], Haar wavelet seems to be the most appropriate. Indeed, its simple mathematical formulation reduces computation time for an online application. Moreover, the decomposition of a signal by Haar wavelet yields constant values over constant time intervals, which may be interesting for a low-frequency signal.

Level of decomposition:

A priori knowledge and expertise on energy generation and/or storage enables the identification of energy sources that can accept frequency ranges of power profiles without altering their operation, which minimizes their degradation. A first estimation of the level of signal decomposition can be done according to the following data. Indeed, according to [AGB 11, AKL 08, BOU 09], the frequency ranges of the FC, batteries and ultracapacitors are included in the following intervals:

– FC: $[10^{-4}, 10^{-2}]$ Hz;

– batteries: $[10^{-2}, 10^{2}]$ Hz;

– UC: $[10^{-2}, 10^{6}]$ Hz.

Therefore, it seems obvious that low-frequency signals should be assigned to FC and batteries, and high-frequency signals to UCs. In wavelet decomposition, the approximation signal always contains low frequencies. In other terms, it contains frequencies that are less than or equal to a certain frequency, f_c.

Considering a numerical signal S with its sampling frequency f_s and n, the chosen level of decomposition, the following is obtained:

$$f_c = \frac{Nf}{2^n} = \frac{f_s}{2^{n+1}} \qquad [3.10]$$

where Nf is the Nyquist frequency that amounts to $\frac{f_s}{2}$ [DOU 05, WAD 10, WIL 08].

The approximation signal A contains the following frequencies: $[0, f_c] = \left[0, \frac{f_s}{2^{n+1}}\right]$ while the detail signal D contains the frequencies comprised in the following interval: $\left[\frac{f_s}{2^{n+1}}, \frac{f_s}{2^n}\right]$.

According to equation [3.10], the decomposition level n should verify the following condition:

$$n = \left\lceil \frac{\log \frac{f_s}{f_c}}{\log 2} - 1 \right\rceil \qquad [3.11]$$

Next, a presentation of the wavelet transform applied to ECCE vehicle is given, considering various combinations of architectures with various parameterizations depending on the energy sources used. Two combinations have been retained here. The first involves the association of the FC with UCs, and the second is composed of a FC, UC and batteries.

Results obtained for the FC/UC hybridization:

The frequency ranges for optimal use of the FC and UCs have already been indicated. Besides the high-frequency signals intended for the UCs, the negative powers of the signal corresponding to the recovery of braking energy need to be absorbed here. Indeed, the FC is a unidirectional generator.

For a numerical application of the wavelet transform, a first requirement is to set an optimal value of the cutoff frequency for the FC. A frequency f_c = 0.01 Hz is retained here, and the sampling frequency is set at f_s = 100 Hz. Using equation [3.11], the decomposition level is equal to $n = 12$.

In Figure 3.8, a 12-level decomposition yields the approximation signal $A12(n)$. The positive part constitutes the power profile required by the FC (P_{FC}). The detail signals obtained by WT ($D1(n)$... $D12(n)$) and the addition of the negative part of $A12(n)$ are regrouped to form the power profile required by UCs (P_{UC}). The various power profiles are defined by the following relations:

$$\begin{aligned} &P_{PAC} = A12(n) if\ A12(n) \geq 0 \\ &P_{PAC} = 0 \qquad if\ A12(n) < 0 \end{aligned} \qquad [3.12]$$

$$\begin{aligned} &P_{SC} = D1(n) + \ldots + D12(n) \qquad\qquad if\ A12(n) \geq 0 \\ &P_{SC} = D1(n) + \ldots + D12(n) + A12(n)\ if\ A12(n) < 0 \end{aligned} \qquad [3.13]$$

Results obtained for the FC/UC/Batteries hybridization:

Two frequencies should be considered for this type of architecture: that of the battery and that of the FC. In the previous part, the frequency of the FC has been set at f_{c1} = 0.01 Hz (corresponding to the decomposition level of 12). For the battery, the retained frequency is f_{c2} = 5 Hz corresponding to a decomposition level of 3.

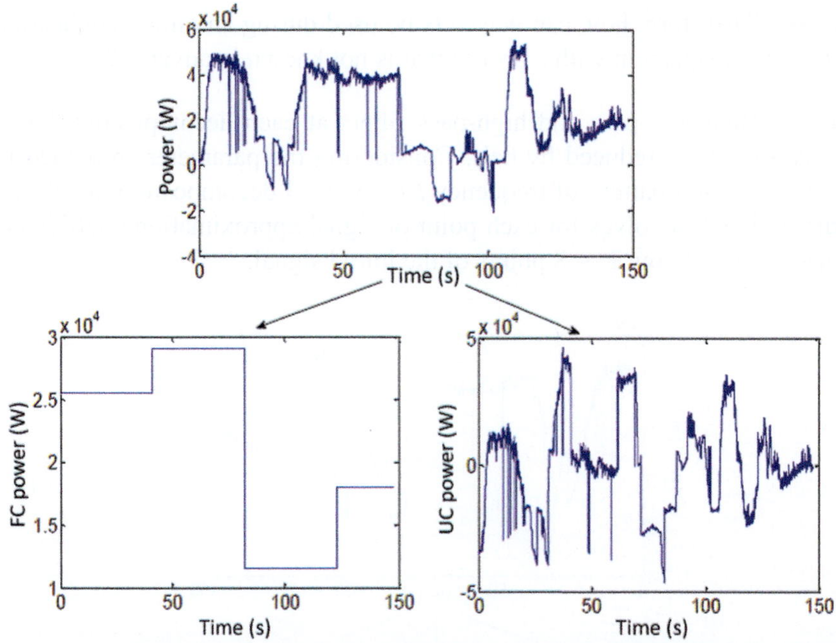

Figure 3.8. *Wavelet decomposition for FC/UC architecture*

For a fuel cell of frequency f_{c1} = 0.01 Hz, the proper decomposition level is n = 12. After elimination of the negative part, the signal A12 is assigned to the FC. Moreover, as it has been shown in the previous section, for a battery of frequency f_{c2} = 5 Hz, the appropriate decomposition level is 3. Power is distributed (Figure 3.9) according to the following relations:

$$\begin{aligned} &P_{PAC} = A12(n) \; if \; A12(n) \geq 0 \\ &P_{PAC} = 0 \qquad if \; A12(n) < 0 \end{aligned} \qquad [3.14]$$

$$\begin{aligned} &P_{Batteries} = D1(n) + D2(n) + D3(n) \qquad if \; A12(n) \geq 0 \\ &P_{Batteries} = D1(n) + D2(n) + D3(n) + A12(n) \; if \; A12(n) < 0 \end{aligned} \qquad [3.15]$$

$$P_{SC} = D4(n) + \ldots + D12(n) \qquad [3.16]$$

In this first study, the wavelet transform has been applied to an *a priori* known power demand (mission profile) signal, which is a strong hypothesis in the case of online management. Indeed, this profile cannot be known in a real case. Therefore, how can wavelets be used during an online application of energy management with a profile that is not known in advance?

Using Haar low-pass and high-pass filters at each decomposition level, the size of data is reduced by half. Considering the parameters provided in this chapter, for a battery of frequency f_{c2} = 5 Hz, 3 decomposition levels are required, which involves for each point of signal approximation coefficients the necessity to know $2^3 = 8$ points of the initial signal.

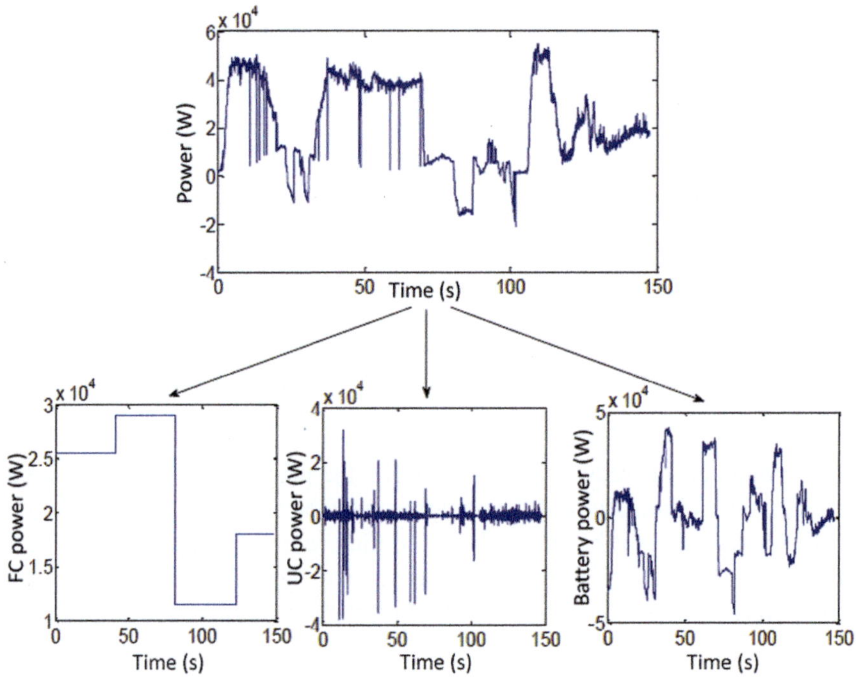

Figure 3.9. *Wavelet decomposition for FC/UC/Batteries architecture*

A prediction algorithm is thus required for a proper distribution of energy among sources. In the next two sections, the predictive aspect is elaborated for online energy management. The mathematical methods used are autoregressive integrated moving average (ARIMA) models and nonlinear autoregressive neural networks (NARNN).

3.3.2. *Application of ARIMA models for energy management*

As noted in the previous section, a time prediction of the studied signal must be done when the wavelet transform is implemented for power distribution in a hybrid vehicle. Many models enable time series prediction. Such examples are vector autoregressive models, state space models, probabilistic models or neural models. Most of the listed models (except for NN) are not suited for this application, as they often require significant computation times [IBR 13].

ARIMA (autoregressive integrated moving average) models are of interest for our work, as they are reliable mathematical models and their formulation is simple. Moreover, they involve short computation times, are able to model short time series (≥ 50 points) and the prediction of one point depends only on the past values of the series itself and not on external data. Furthermore, they have already been tested in applications centered on electric and energy systems. Indeed, in [ISM 05], seasonal ARIMA models (SARIMA) have been used to evaluate the amount of electricity generated in Malaysia, in [KOS 08] they have been used to estimate the low-frequency electromechanical modes in an energy system and in [CHU 13], to analyze the consumption of electricity; the results obtained in these various references have shown small prediction errors, which proves their robustness for such uses.

Referring to batteries, ARIMA models have been implemented in [KAZ 03] to predict their state of health and their state of charge. The prediction error was about 5%. In [SAH 09], ARIMA models were used to predict the lifetime of a battery with small errors over a 32-week horizon.

Consequently, a brief presentation of how the ARIMA model is constructed is given here, followed by the online management algorithm and finally by the results obtained when coupling the ARIMA model with the wavelet transform.

3.3.2.1. *ARIMA model construction*

A process Y_t is referred to as autoregressive integrated moving average of order p, d and q if after d differentiations (difference between successive values of the time series), it can be written in the following form equation [3.17]:

$$Y_t = c + \varphi_1 Y_{t-1} + \varphi_2 Y_{t-2} + \ldots + \varphi_p Y_{t-p} + U_t + \theta_1 U_{t-1} + \theta_2 U_{t-2} + \ldots + \theta_q U_{t-q} \qquad [3.17]$$

where U_t is a white noise, and ϕ_i and θ_j are real numbers.

This ARIMA (p, d, q) model is a combination of two models MA(q) (moving average of order q) and AR(p) (autoregressive model of order p). Moreover, this model is of stationary type.

An ARIMA model is built in the following three stages, according to the approach of Box–Jenkins [MAK 97, SHU 11]:

– model identification;

– model estimation;

– residue test or diagnosis.

The first step required is model identification. To identify the order of an ARIMA (p, d, q) process, the minimization of the Akaike information criterion [BOX 08] is used. This tool enables the comparison of various models depending on various values of p and q, in order to adjust the same time series. Parameter d is chosen so that the series is stationary.

Parameters ϕ_i and θ_j are estimated with the maximum likelihood method. This method is currently used for the estimation of the coefficients of time series model, as its implementation remains simple.

Finally, the residues between the values of the estimated model and the observed values of the series must behave as a white noise. The Portmanteau test has been chosen for this work, which enables the evaluation of the correlation between residues.

The methods developed for defining the set of these parameters are detailed in the thesis by Mona Ibrahim [IBR 13]. The soundest results of various studies conducted by Mona Ibrahim are alone presented here.

3.3.2.2. *Online management algorithm based on ARIMA models*

This method involves taking part of the vehicle power demand signal, assigning an ARIMA (p, d, q) model and then testing the ability of the model to predict the power signal over a defined horizon. The prediction horizon then defines the maximum decomposition level by the wavelet transform. Indeed, as mentioned in section 1.3.1.2, for a decomposition level j, each coefficient of the approximation and detail signals requires 2^j points of the initial signal. Thus, if the model can predict n points and there is an integer k so that $2^k \leq n$, then it is possible to realize a k-level wavelet decomposition.

To illustrate this methodology, let us consider a real power signal acquired with a sampling frequency of 1 Hz (Figure 3.10). To be as realistic as possible, the first 64 points of this signal are only considered. Indeed, during the vehicle start, the various sources of energy must be able to supply the required power and energy for 64 s. The energy management strategy does not come into play during this time span, and prediction is not necessary. To obtain the best prediction results, wavelet denoising is applied to this signal. The aim is to rebuild a version of a given signal, after having eliminated the noises according to thresholding [MAL 08, MIS 07b].

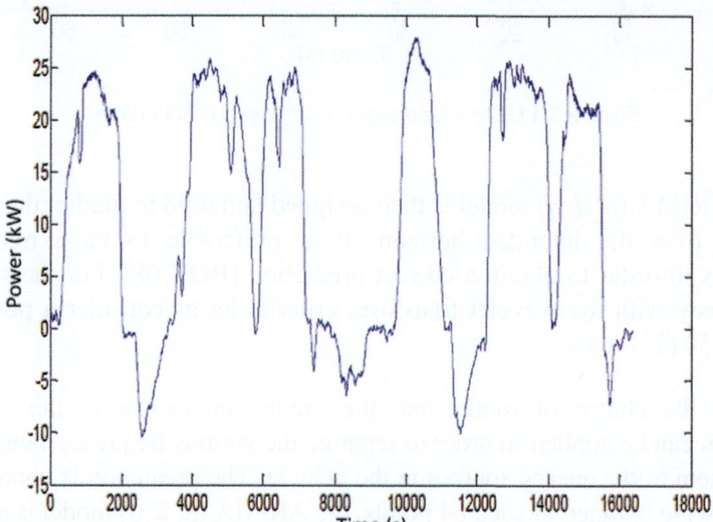

Figure 3.10. *Mission profile (sampling frequency of 1 Hz)*

The signal obtained after denoising is shown in Figure 3.11. It has the advantage of showing a soft evolution, without sudden variations.

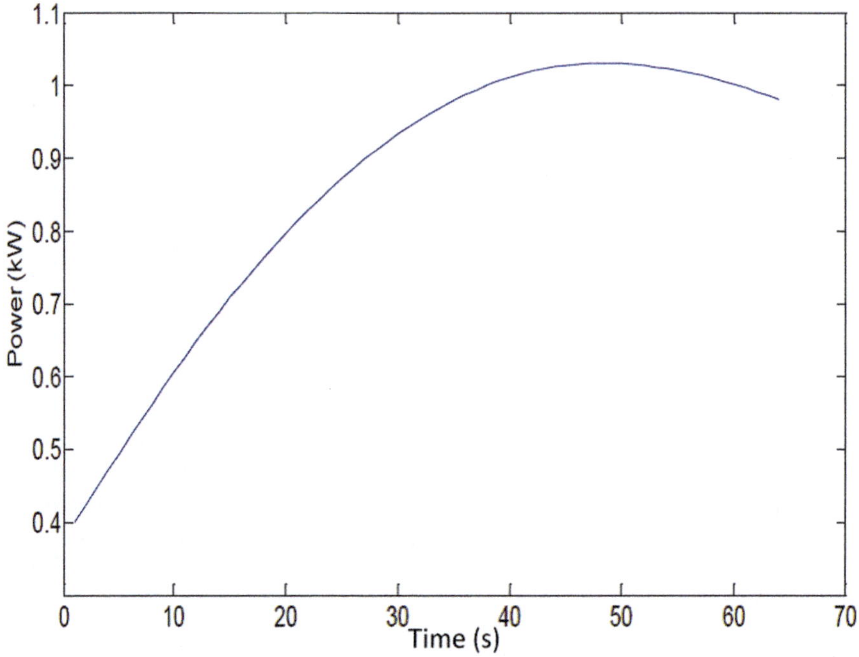

Figure 3.11. *Denoised signal composed of 64 points*

An ARIMA (p, d, q) model is then assigned and used to predict the power demand over the intended horizon. It is preferable to have minimum 50 points in order to obtain a correct prediction [BOX 08]. For the sake of consistency with the wavelet transform criteria, let us consider a power of 2 above 50 ($2^6 = 64$).

Once the choice of model and the prediction are made, the wavelet transform can be applied in order to separate the various frequency ranges and assign them to the energy sources in the vehicle. The prediction is obtained in the following manner: at each 64 points, the ARIMA (3, 2, 0) model is applied to predict 64 new points. The advantage of this online algorithm is that real data are available at each new prediction of 64 points. The process is renewed throughout the mission duration (Figure 3.12(a)). Following the prediction, the six-level wavelet transform is activated for the real-time power distribution to each source within their corresponding frequency range.

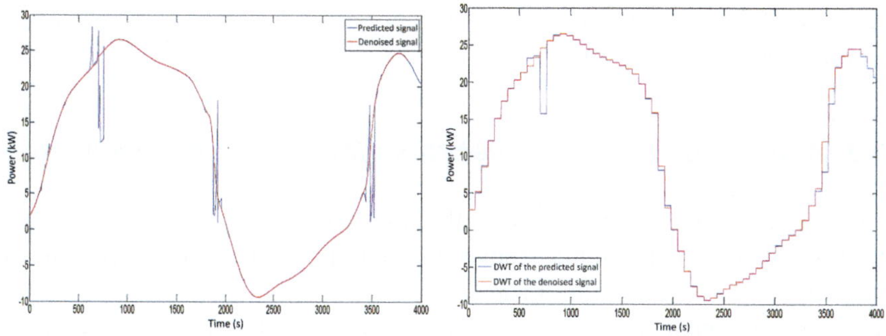

a) b)

Figure 3.12. *Prediction of the signal a) and its DWT b), with a horizon of 64 points. For a color version of this figure, see www.iste.co.uk/jemei/cells.zip*

Real-time performance of these operations requires a consideration of the computation times. The prediction time using ARIMA model is 0.8 s, which is shorter than the sampling interval set here at 1 s. The error evidenced here is 3 W^2. As these values indicate, this algorithm can be used in real-time.

The results obtained for the approximation signal are presented in Figure 3.12(b). The frequency of this signal ranges between 10^{-3} and 10^2 Hz.

Thus, thanks to the frequency ranges obtained in the approximation signal, it is possible to distribute the power to the FC and the batteries. Indeed, let us recall that the frequency range within which the FC operates in proper conditions is between 10^{-4} and 10^{-2} Hz and that of the batteries is between 10^{-2} and 10^2 Hz. The detail signals, representing the high-frequency power are oriented towards ultracapacitors.

It is worth recalling that the main purpose of using the wavelet transform in this study is to increase the lifetime of the FC and other components. No study or measure intended for this purpose has been conducted. Nevertheless, given that the FC operates in the optimal frequency range, it can be concluded that the degradations are not due to dynamic stresses of the mission profile. It can therefore be estimated that this method contributes to the increase in the lifetimes of various components, and particularly of the FC.

Moreover, it is worth noting that the computation times required for the execution of various parts of this algorithm (wavelet denoising, ARIMA prediction and the wavelet transform) are shorter than (0.9 s) the sampling time (1 s). Consequently, this opens the possibility of using this algorithm for online energy management.

3.3.3. *Applications of neural networks for energy management*

The ARIMA models presented in the previous sections are linear models. The results obtained remain consistent, although the time series to be modeled in our work are nonlinear. In view of obtaining higher-performance energy management, neural networks will be considered, with a focus on nonlinear autoregressive neural networks (NARNN). Proof that a time series can still be modeled by a nonlinear autoregressive (NAR) model can be found in [LAP 87]. This is described by equation [3.18]:

$$y(t) = h\big(y(t-1), y(t-2), \cdots, y(t-p)\big) + \varepsilon(t) \quad [3.18]$$

An NARNN (nonlinear autoregressive neural network) is a discrete nonlinear and autoregressive model with endogenous inputs. Its formulation is given by equation [3.19]:

$$\hat{y}(t) = h\big(y(t-1), y(t-2), \cdots, y(t-p)\big) + \varepsilon(t) \quad [3.19]$$

It is a recurrent and dynamic multilayer network, whose loops are illustrated in Figure 3.13 [ALL 11]:

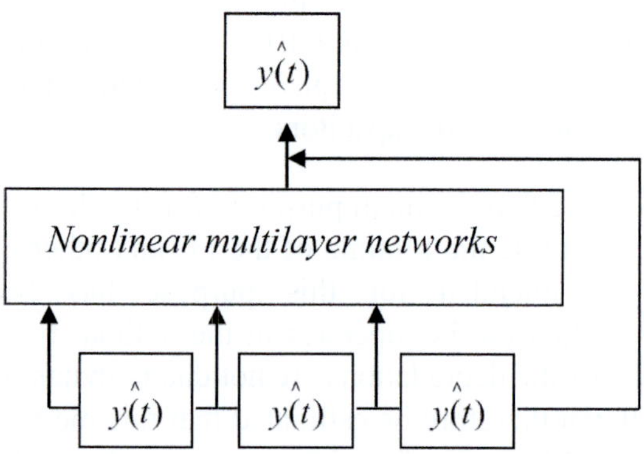

Figure 3.13. *Structure of a recurrent NARNN*

In such a structure, learning relies on the decreasing of the propagation algorithm's gradient. NARNN has been shown in this work because it allows us to approach the problems that have long time dependence, which is not the case with other classic recurrent neural networks (Elman, Jordan, Hopfield, etc.). Indeed, as shown in [BEN 94, LIN 96], learning cannot be successful on databases that contain information covering a long period of time. Hence, NARNN is often used for modeling and predicting time series.

3.3.3.1. *NARNN parameterizing*

The methodology used for developing the ARIMA-based energy management algorithm is resumed here. To compare the results of the two methodologies, the same input data are used for the prediction, namely the first 64 points of the denoised signal (Figure 3.11).

However, in order to predict the data, NARNN needs to be parameterized.

The set of points ($y(t-1), y(t-2),..., y(t-p)$) of equation [3.19] forms what is commonly known as the time window. One way of choosing p is to go through the representation of the partial autocorrelation function (PACF), which expresses the link between y_t and $y_{(t-k)}$ for $k = 1,..., p$ when the influence of intermediate values $y_{(t-k+1)},..., y_{(t-1)}$ is withdrawn. This method makes it possible to set $p = 3$. Thus, the number of NARNN inputs is 4 [IBR 13, IBR 15a, IBR 15b] ($y_t, y_{t-1}, y_{t-2}, y_{t-3}$).

Among the elements to be parameterized, the number of hidden layers and the number of neurons on each layer are dominant factors for the performances of a neural network. In order to define these parameters, several architectures have been tested and the one offering the best compromise between the learning time and the medium square error (MSE) has been retained. This involves a hidden layer with 10 neurons. In order to test this architecture, a signal with the sampling frequency of 10 Hz has been used (Figure 3.14). Let us note that the algorithm diverges after the first 10 points and that the prediction is no longer correct.

An initial comparison between NARNN and ARIMA can be drawn. Indeed, the results obtained with the ARIMA model show that it is possible to make a prediction over more distant horizons than that obtained with NARNNs. Nevertheless, neural networks have higher execution speed, which remains a significant advantage for real-time implementation. In fact, neural networks seem to be the adequate tool for short-term prediction.

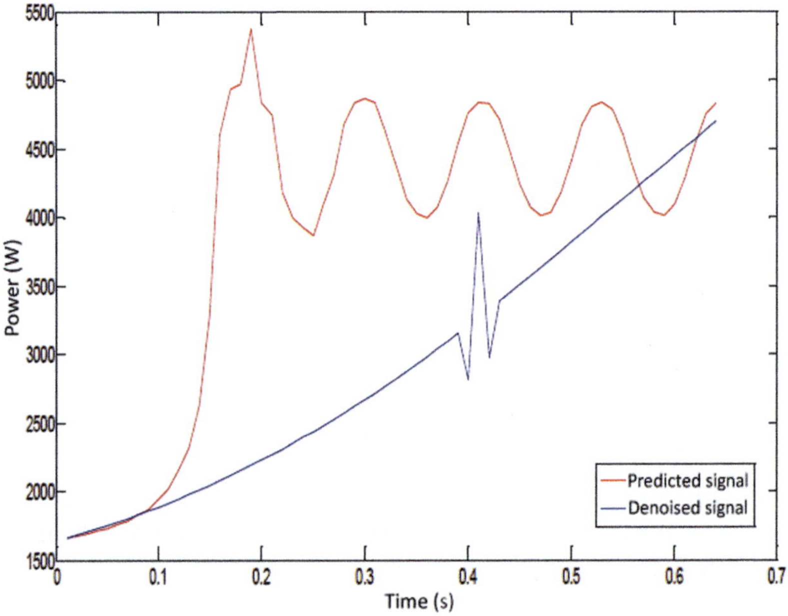

Figure 3.14. *Results of prediction using NARNN. For a color version of this figure, see www.iste.co.uk/jemei/cells.zip*

It is worth keeping in mind that the models used (ARIMA and NARNN) have been parameterized on the same learning basis (first 64 points of a denoised signal) to realize predictions. In order to improve the results obtained, it is important to test the adaptive models that are parameterized after each prediction horizon. Indeed, thanks to new available data, it is possible to obtain a new set of parameters that should enable a more precise prediction. This will be the focus of the next section.

3.3.4. *Comparison of adaptive NARNN and ARIMA*

3.3.4.1. *Adaptive NARNN*

The same time series will be used here for testing adaptive NARNN. After a first prediction over a horizon *h*, the model achieves learning with the new real inputs available in order to adapt the parameters. In other terms, a sliding window is involved. The prediction results for the 10 Hz signal are presented in Figure 3.15. The prediction horizon is set here at 20 points, and the error obtained is 3×10^{-5} W^2. The resulting wavelet transform that makes it possible to obtain the approximation signal to be sent to the FC is perfectly adequate (Figure 3.16).

3.3.4.2. *Adaptive ARIMA*

Similar to the adaptive NARNN, the ARIMA (p, d, q) model is reevaluated after each prediction horizon. Let us recall that the prediction horizon is now 64 points. The results obtained for the prediction on a 10 Hz signal are represented in Figure 3.17, and the resulting wavelet transform is shown in Figure 3.18. The error noted on the prediction signal is 7×10^{-5} W^2.

The errors obtained with the adaptive ARIMA model are very small. It allows perfectly adequate prediction horizon compared to adaptive NARNN (64 compared to 20 points). Furthermore, execution times shorter than 1 s enable the consideration of a real-time implementation for signals with sampling frequency less than or equal to 1 Hz.

In summary, the adaptive models provide better prediction results. It is worth noting that they require very few learning points (20 for NARNN and 64 for ARIMA) to provide a reliable prediction for the entire signal. The computation times of adaptive networks are necessarily longer than those of non-adaptive models, but they nevertheless remain quite acceptable for an online application. Finally, the adaptive ARIMA models appear to show better performances than neural networks, especially in terms of prediction errors [IBR 16].

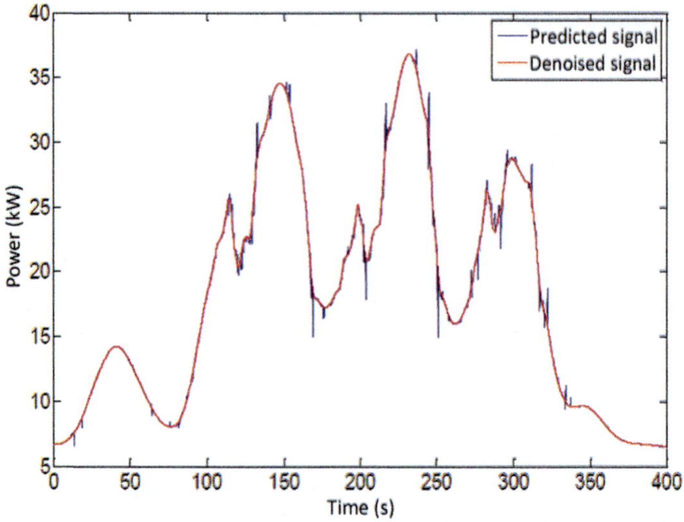

Figure 3.15. *Results of prediction with adaptive NARNN (10 Hz signal). For a color version of this figure, see www.iste.co.uk/jemei/cells.zip*

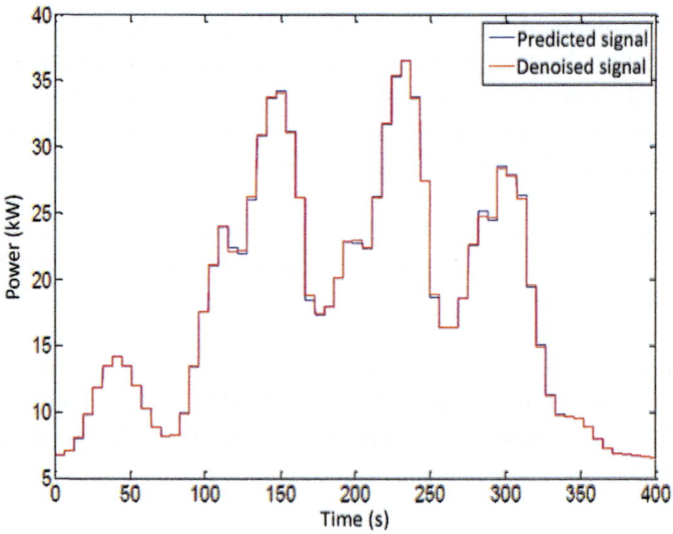

Figure 3.16. *Approximation signal resulting from DWT (NARNN – 10 Hz signal). For a color version of this figure, see www.iste.co.uk/jemei/cells.zip*

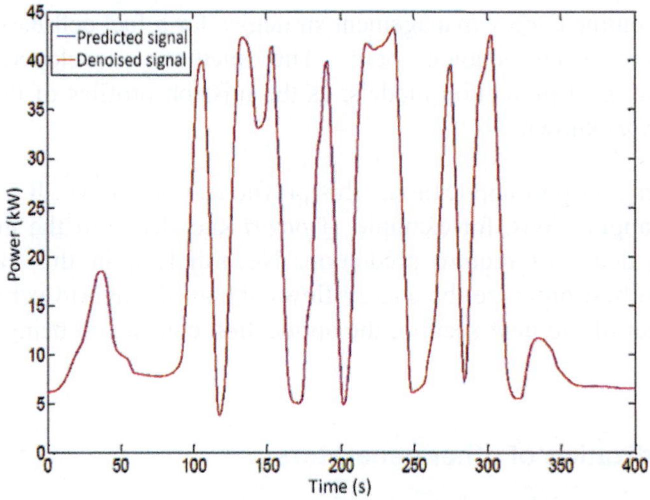

Figure 3.17. *Results of prediction with adaptive ARIMA (10 Hz signal). For a color version of this figure, see www.iste.co.uk/jemei/cells.zip*

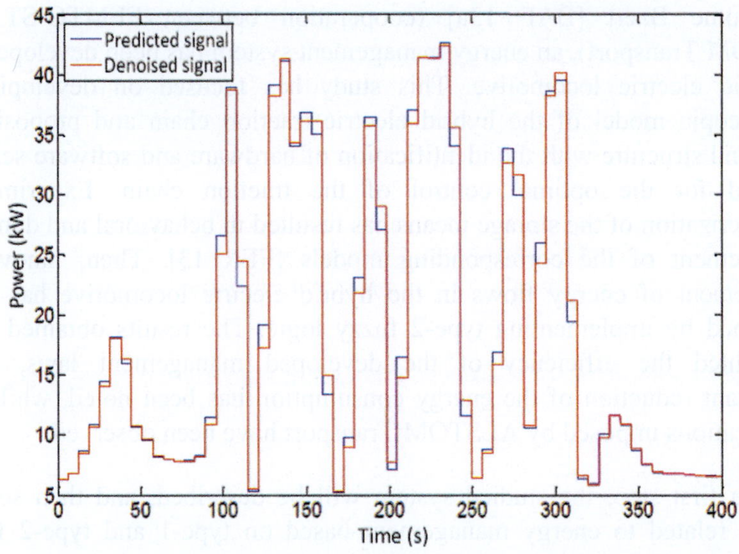

Figure 3.18. *Approximation signal resulting from DWT (ARIMA – 10 Hz signal). For a color version of this figure, see www.iste.co.uk/jemei/cells.zip*

Various online energy management strategies for a fuel cell-based hybrid vehicle have been studied here. The methods developed require implementation of prediction models, as the mission profiles of the vehicle are not *a priori* known.

For other transport applications, this profile can be known. It is the case of railway applications, for example. *A priori* knowledge of the mission to accomplish does not require prediction. Nevertheless, in this case, it is important to best optimize the energy flows aboard the hybrid vehicle. This is the subject of the next section, the application considered being a hybrid locomotive.

3.4. Hybridization of other generators

Power distribution in a hybridized system can be performed offline or online depending on the application and on whether or not the mission profile is *a priori* known. For example, in railway goods transport, the mission profile can be known in advance. Thus, in the context of the thesis by Jérôme Baert [BAE 13a] (cooperation between FEMTO-ST and ALSTOM Transport), an energy management system has been developed for a hybrid electric locomotive. This study has focused on developing a macroscopic model of the hybrid electric traction chain and proposing a command structure with the identification of hardware and software sensors required for the optimal control of the traction chain. Experimental characterization of the storage means has resulted in behavioral and dynamic improvement of the corresponding models [FER 13]. Then, innovating management of energy flows in the hybrid electric locomotive has been developed by implementing type-2 fuzzy logic. The results obtained have highlighted the efficiency of the developed management laws, as a significant reduction of the energy consumption has been noted, while the specifications imposed by ALSTOM Transport have been observed.

As a first step, the studied system will be described, and then several notions related to energy management based on type-1 and type-2 fuzzy controllers will be introduced. Finally, the retained strategy and the results obtained will be presented.

3.4.1. *Topology of the studied system and its modeling*

The activities of the local FRET are generally conducted by diesel locomotives that are both noisy and polluting. In order to obtain less polluting FRET transport solutions, Alstom Transport proposes a solution aimed to couple the traditional diesel propulsion with a more environmentally friendly system of batteries/ultracapacitor packs (Figure 3.19). In such a configuration, the generator has two functions: recharging batteries and ultracapacitors. It provides an energy reserve for successful completion of the important acceleration phases of the locomotive. It also proves that the engine generally operates at low load. Coupling the batteries and the ultracapacitors, this primary source could operate for more significant loads, which would also improve its efficiency. The topology of such a hybrid locomotive is represented in Figure 3.20 [BAE 13b]. This system is composed of the following:

– Ultracapacitors: this reversible source enables the supply and absorption of significant powers with rapid dynamics. Due to its slow internal resistance and its low voltage, it is a rapidly chargeable/dischargeable element with high efficiency.

– Electrochemical accumulators: they are intended for storing electric energy that can be used subsequently. Their operation relies on electrochemical reactions at their electrodes, which ensure the conversion of electric energy into a reversible chemical process. This source has a significant mass energy and stores energy, while maintaining the system's slow dynamics.

– Generator [BAE 12a]: it is the primary source of the locomotive and an autonomous device that can generate electricity. Most generators are composed of an internal combustion engine that acts as an alternator. The more powerful units are driven by gas turbines or high-power diesel engines. The autonomy of the vehicle is only limited by the size of its fuel tank. Moreover, it does not allow energy recovery and, with the electrochemical accumulators, it ensures the slow dynamics required by the locomotive.

– Braking rheostat: they are very widely used when the supply network is not suited for recovery braking and it is naturally used in cases of autonomous tractions. There are two such main types: natural convection rheostats and forced ventilation rheostats. In other terms, the braking rheostat allows for the dissipation of the energy surplus, if needed.

– Central bus capacitor: it is a capacitor that helps us to impose a so-called "bus" voltage common to all the sources and elements onboard the locomotive.

The difficulty posed by using this system is to be able to manage the power exchanges between the existing sources depending on their physical state and on that of the hybrid electric locomotive. The definition and development of an energy management strategy is therefore required.

Nevertheless, before developing this strategy, a dynamic modeling of each of the elements is required, as well as the elaboration of the local control structures. For this purpose, an energetic macroscopic representation (EMR) has been implemented. EMR is a synthetic representation of complex multiphysical systems. This formalism, based on integral causality, allows us to observe the power exchanges between process subsystems. Moreover, model inversion (based on integral causality) facilitates the definition without difficulty of the maximum and/or practical command structure (MCS and/or PCS), while determining the location of hardware and software sensors. The EMR of the hybrid locomotive is shown in Figure 3.21.

None of these models (EMR, MCS or PCS) will be developed here. Interested readers can refer to the article by [BAE 14].

Figure 3.19. *Hybrid electric locomotive. For a color version of this figure, see www.iste.co.uk/jemei/cells.zip*

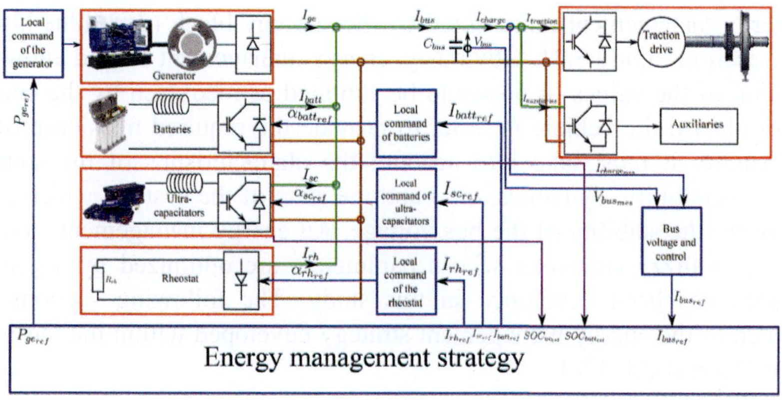

Figure 3.20. *Architecture of the hybrid locomotive. For a color version of this figure, see www.iste.co.uk/jemei/cells.zip*

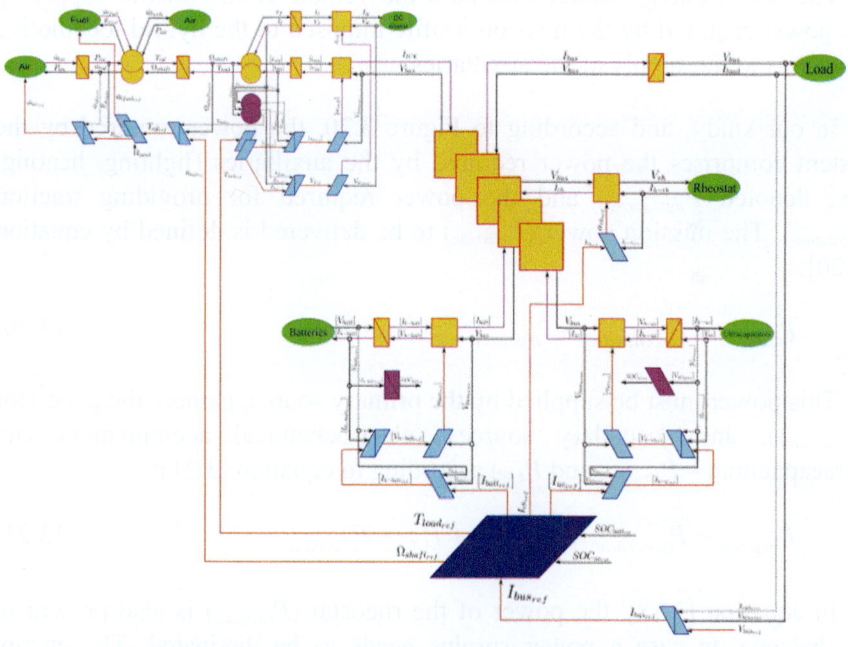

Figure 3.21. *EMR architecture of the hybrid locomotive. For a color version of this figure, see www.iste.co.uk/jemei/cells.zip*

Energy management represented by the strategy block (dark blue block) can be seen in Figure 3.21. Obviously, energy management should allow the definition of the values of power to be supplied or recovered by the energy sources aboard the vehicle in order to provide the required mission profile. Furthermore, it must take into account the characteristics of the sources (SOC, reversibility, dynamics, etc.) and the state of the system, which must also ensure the stability of the bus voltage. An energy management strategy based on a fuzzy controller whose parameters are optimized by a genetic algorithm has been developed in this study. The following sections are dedicated to the energy management strategy developed within the thesis by Jérôme Baert [BAE 13a].

3.4.2. *Energy management strategy*

The set of energy sources onboard the vehicle contributes to supplying the power required by the mission profile imposed to the hybrid locomotive as well as to the supply of the auxiliaries.

In our study, and according to Figure 3.20, the power required by the system comprises the power required by the auxiliaries (lighting, heating, etc., denoted $P_{auxiliaries}$) and the power required for providing traction ($P_{traction}$). The mission power ($P_{mission}$) to be delivered is defined by equation [3.20]:

$$P_{mission} = P_{auxiliaries} + P_{traction} \qquad [3.20]$$

This power must be supplied by the primary source, namely the generator ($P_{generator}$), and secondary sources (electrochemical accumulators and ultracapacitors – $P_{batteries}$ and P_{UC}) according to equation [3.21]:

$$P_{mission} = P_{generator} + P_{batteries} + P_{UC} + P_{rheostat} \qquad [3.21]$$

In equation [3.18], the power of the rheostat ($P_{rheostat}$) is also present in the balance, in case a power surplus needs to be dissipated. The energy management strategy is an algorithm that controls the power distribution in the hybrid electric locomotive in an optimized manner. The implemented energy management strategy defines the reference states of charge of the batteries and the ultracapacitors, respectively, depending on the acceleration and speed of the locomotive (equations [3.22] and [3.23]) [BAE 12b]:

$$SOC_{Batt\cdot ref} = SOC_{Batt\cdot \max} - \left|\frac{\dot{v}_{veh}}{\dot{v}_{\max}}\right|\left(SOC_{Batt\cdot \max} - SOC_{Batt\cdot \min}\right) \qquad [3.22]$$

$$SOC_{SCAP_{ref}} = SOC_{SCAP_{\max}} - \left|\frac{v_{veh}}{v_{\max}}\right|\left(SOC_{SCAP_{\max}} - SOC_{SCAP_{\min}}\right) \qquad [3.23]$$

where $SOC_{x\text{-}ref}$ is the reference state of charge, $SOC_{x\text{-}max}$ and $SOC_{x\text{-}min}$ are the states of maximum and minimum charge admitted by the considered element, respectively, v_{veh} is the speed ($\dot{v}_{veh}$ is the acceleration) of the vehicle and v_{max} ($\dot{v}_{\max}$ is the maximum acceleration) is the maximum speed of the vehicle. x denotes the case Batt. for the batteries or UC for the ultracapacitors. It is preferable to use the acceleration of the vehicle to enable the reference state of charge of the batteries to be as high as possible. Moreover, two limitation factors are used to avoid exceeding these states of minimum and maximum charge. Finally, it should be noted that the power of the mission is separated into high and low frequencies through the intermediary of a low-pass filter. The high-frequency powers are supplied by the ultracapacitors, and the low-frequency powers are supplied by both the batteries and the generator in order to increase the lifetime of the batteries and avoid the dynamic stress of the generator.

Fuzzy logic is used for the completion of this energy management strategy. This method belongs to the category of so-called online techniques. The algorithms in this category operate in real-time, but the solutions they advance are not necessarily optimal. Their principle is to operate an optimization considering the studied system in its entirety and neglecting certain constraints in order to ease the optimization process. Unlike the offline techniques, they do not require *a priori* knowledge of the mission profile. Such methods also include the neural networks, predictive control and the wavelet transform. Here, the use of fuzzy logic enables the development of a solution in which the mission profile is not known *a priori*, and the real driving conditions are taken into account by relying on information such as speed, acceleration, state of charge and power calls.

The developed fuzzy controller (Figure 3.21) has two inputs and one output [SOL 12a]. The first input (*dP*) corresponds to the difference between the power required by the mission and the power delivered at the same instant by the generator equation [3.24]. Dividing this difference by the maximum power required by the mission enables the normalization of this input. It is aimed to provide the main energy supply to the vehicle:

$$dP = \frac{P_{mission} - P_{GE_{mes}}}{P_{mission_{\max}}} \quad [3.24]$$

The second input (e_{batt}) is the difference between the estimated state of charge of the batteries and a reference state of charge defined by equation [3.22]. It aims to maintain the state of charge of the batteries. The output of the fuzzy controller (dP_{GE}) is the relative variation of power applied to the generator. Therefore, it enables the increase or decrease in the power supplied to the system. The controller is shown in Figure 3.22. For generalization reasons, the inputs and the output of the controller are normalized between [−1;+1]. The affiliation functions used are triangles or trapezoids and their number is seven per input/output. Then, the inference table (Table 3.1) counts 49 rules.

***dP* ↓ *ebatt* ➔**	***NH***	***NM***	***NL***	***Z***	***PL***	***PM***	***PH***
NH	*NH*	*NH*	*NH*	*NM*	*NM*	*NL*	*Z*
NM	*NH*	*NH*	*NM*	*NM*	*NL*	*Z*	*PL*
NL	*NH*	*NM*	*NM*	*NL*	*Z*	*PL*	*PM*
Z	*NM*	*NM*	*NL*	*Z*	*PL*	*PM*	*PM*
PL	*NM*	*PN*	*Z*	*PL*	*PM*	*PM*	*PH*
PM	*PN*	*Z*	*PP*	*PM*	*PM*	*PH*	*PH*
PH	*Z*	*PL*	*PM*	*PM*	*PH*	*PH*	*PH*

Table 3.1. *Inference table*

With “NH – Negative High”, “NM – Negative Medium”, “NL – Negative Low”, “Z – Zero”, “PL – Positive Low”, “PM – Positive Medium”, “PH – Positive High”.

Example: IF dP is PH AND e_{batt} is PH, THEN dP_{GE} is PH. IF the power delivered by the generator is much less than the power required by the mission AND if the batteries must be "highly" recharged, THEN the power delivered by the generator is significantly increased.

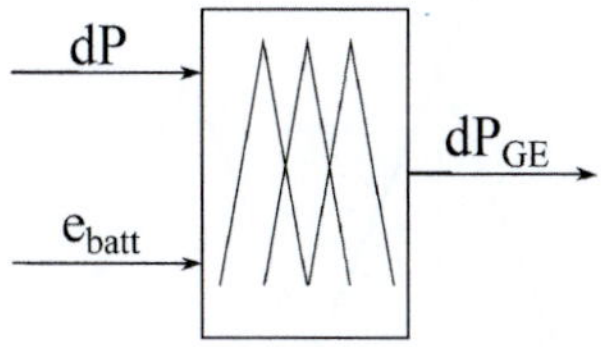

Figure 3.22. *Structure of the fuzzy controller*

3.4.3. *Type-2 fuzzy logic and optimization of the fuzzy controller*

Very conclusive works have been conducted by Javier Solano [SOL 11, SOL 12a, SOL 12b] within our team. Indeed, an energy management strategy based on type-2 fuzzy logic has been developed and implemented in real-time on an army vehicle of 14 tons (ECCE). Hence, the decision has been made in our team to continue our study using fuzzy logic.

3.4.3.1. *Type-2 fuzzy logic*

The concept of type-2 fuzzy logic sets was introduced by [JOH 07, ZAD 65] as an extension of the concept of ordinary fuzzy set called type-1 fuzzy set. A type-2 fuzzy set is characterized by a fuzzy affiliation function. The value of affiliation of the elements of the set is [0; 1] interval. Type-1 intervenes when it is difficult or even impossible to determine these degrees of affiliation of an element to a set by 0 or 1 (Figure 3.23). Similarly, type-2 intervenes when determining the affiliation fuzzy functions by real numbers within [0; 1] proves complicated (Figure 3.24) [WAG 10].

The parameters of various affiliation functions should be determined. The controller shown in Figure 3.21 uses two inputs and one output associated with seven affiliation functions (Table 3.1). Thus, 21 affiliation functions need to be parameterized. To accomplish this task, a genetic algorithm has been developed for optimizing the position of these affiliation functions.

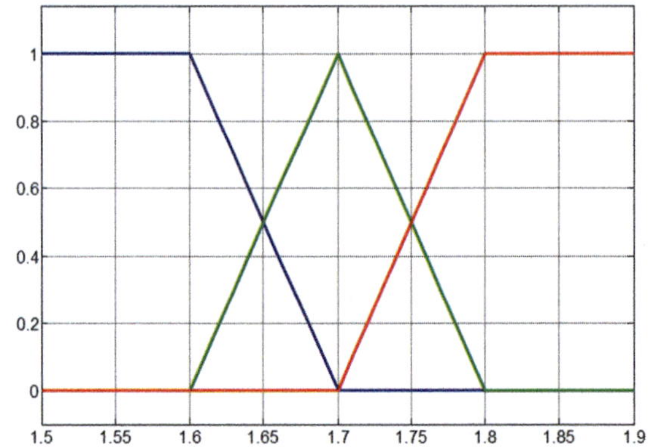

Figure 3.23. *Type-1 fuzzy set. For a color version of this figure, see www.iste.co.uk/jemei/cells.zip*

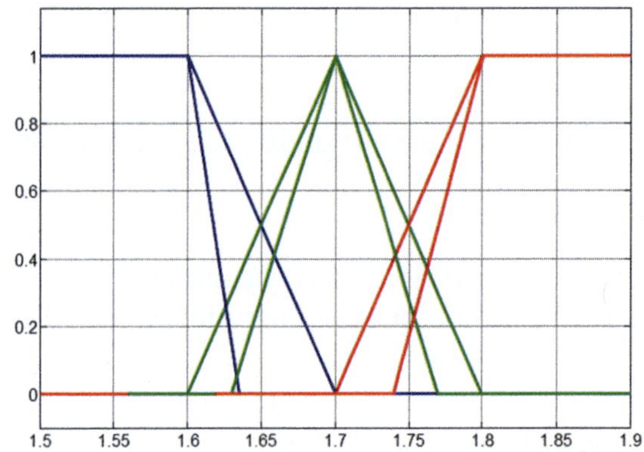

Figure 3.24. *Type-2 fuzzy set. For a color version of this figure, see www.iste.co.uk/jemei/cells.zip*

3.4.3.2. *Genetic algorithm*

The parameters of a fuzzy controller are defined by a genetic algorithm [CHA 07]. Its characteristic is that it draws inspiration from the evolution of species in their natural context. Species adapt to their life environment, which can evolve, individuals of each species reproduce and generate new individuals, some of them undergo modifications of their DNA and others disappear. A genetic algorithm replicates this evolution model in order to find solutions to a given problem. The terms used are therefore borrowed from the real world of biology and genetics. In the present case, a population is a set of individuals (set of parameters of the affiliation functions to be optimized). Therefore, an individual is a solution to a given problem, which is here the minimization of the generator use. A gene is a part of a solution, therefore of an individual. A generation is an iteration of the algorithm. A genetic algorithm drives the evolution of a population with the aim of improving individuals. With each generation, a set of individuals is highlighted. The stages of the developed algorithm are the following (Figure 3.25):

– Chromosomes or individuals are randomly generated. The corresponding fuzzy controllers are synthesized and the values of the fitness function are calculated for each of these controllers. The fitness function should be properly defined, as it allows for the evaluation of each individual's performance with respect to a given problem. In this study, this function is directly related to minimizing the fuel consumption of the hybrid electric locomotive, improving its autonomy and reducing its pollutant emissions. Therefore, this function minimizes the use of the primary source (generator) and maximizes the use of secondary sources (batteries and ultracapacitors) equation [3.25]:

$$\max\left(\int_0^{t_{mission}} \left|P_{Batt.}(t)\right| dt\right) + \max\left(\int_0^{t_{mission}} \left|P_{SCAP}(t)\right| dt\right) + \min\left(\int_0^{t_{mission}} P_{GE}(t)\, dt\right) \qquad [3.25]$$

– The stage of constraints involves verifying that the batteries and ultracapacitors are charged by the end of mission. In such a case, the algorithm goes to the next stage, otherwise, the corresponding individual is eliminated and a new one is synthesized and then tested. The precision used is 0.1%, which means, for example, that if the state of charge of the batteries by the end of mission is below 89.9% (the authorized maximum in the energy management strategy: 90%), then the individual is not validated. Penalties could have been introduced on the individuals, depending on the result obtained at this final state of charge. However, such penalties would degrade the results obtained at the level of fitness function. Indeed, the latter aims to minimize the use of the generator, or batteries can only be recharged by the generator (Figure 3.30). Penalizing "improper recharge" of batteries amounts to penalizing "proper use" of the generator.

– The selection stage uses a biased method per tournament. It involves random and two-by-two selection of the chromosomes or individuals of the initial population. Their respective values of the fitness function are compared. The one with the highest value is selected and the other one is eliminated. In order to maintain and facilitate the diversity of individuals, the selection is biased. The result of the selection stage is a population said selected, whose size is half that of the initial population.

– Crossover at present enables the recombination of the genetic characteristics of the selected population [WAG 09]. In mathematical terms, the selected individuals are coupled in pairs and generate two children. This makes it possible to obtain a final population of the size of the initial population. If $[a_i, b_i, c_i]$ and $[a_j, b_j, c_j]$ are two parents and p is a random variable in the interval [0,1], then two child chromosomes are synthesized (equations [3.26] and [3.27]):

$$child_1 = p \times [a_i, b_i, c_i] + (1-p) \times [a_j, b_j, c_j] \quad [3.26]$$

$$child_2 = p \times [a_j, b_j, c_j] + (1-p) \times [a_i, b_i, c_i] \quad [3.27]$$

– Finally, the mutation modifies with a very low probability (between 0.001 and 0.01) the value of one or several values of the genes of individuals. Diversity and noise are then introduced into the population itself and ensure reaching an overall maximum, in the case of an infinite number of iterations.

This last population is saved as the generation and the previous stages are repeated in order to yield future generations of individuals.

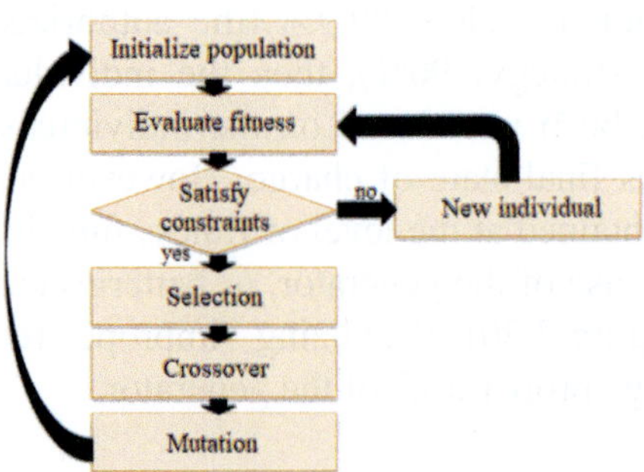

Figure 3.25. *Structure of the developed genetic algorithm*

3.4.4. *Simulation results*

3.4.4.1. *Genetic algorithm*

Figures 3.26–3.28 show affiliation functions obtained by the genetic algorithm. The set of member functions (inputs and output), of which there are 21, required the following:

– for the two inputs (*dP* – Figure 3.26 and e_{batt} – Figure 3.27), 35 parameters each to define their respective 7 affiliation functions (2 inputs × 7 affiliation functions × 5 parameters per function);

– for the output (dP_{GE} – Figure 3.28), 14 parameters were needed for the definition of 7 affiliation functions (1 output × 7 affiliation functions × 2 parameters per function).

Thus constituted, the individuals of the genetic algorithm have a size of 84. Figure 3.29 represents the generated optimum type-2 fuzzy controller. The output of the controller (*dPGE*) is represented depending on the two entries (*dP* and *ebatt*). This fuzzy controller is loaded in the energy management strategy of the studied system and then tested.

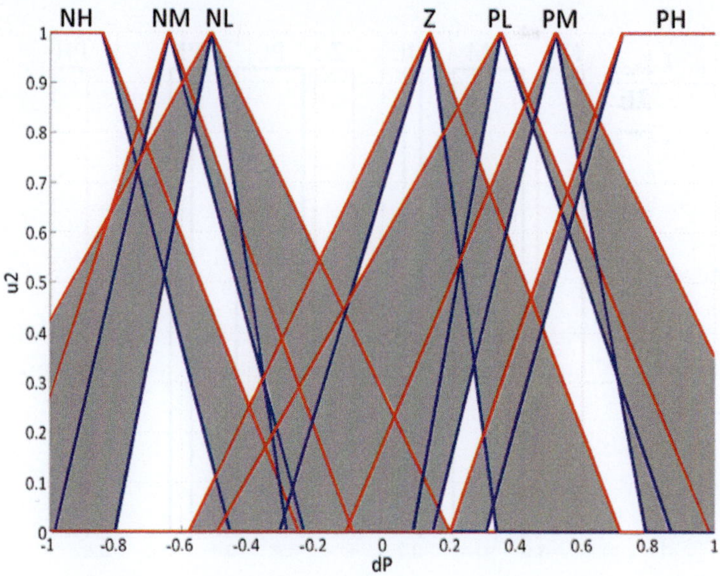

Figure 3.26. *Affiliation functions of input 1. For a color version of this figure, see www.iste.co.uk/jemei/cells.zip*

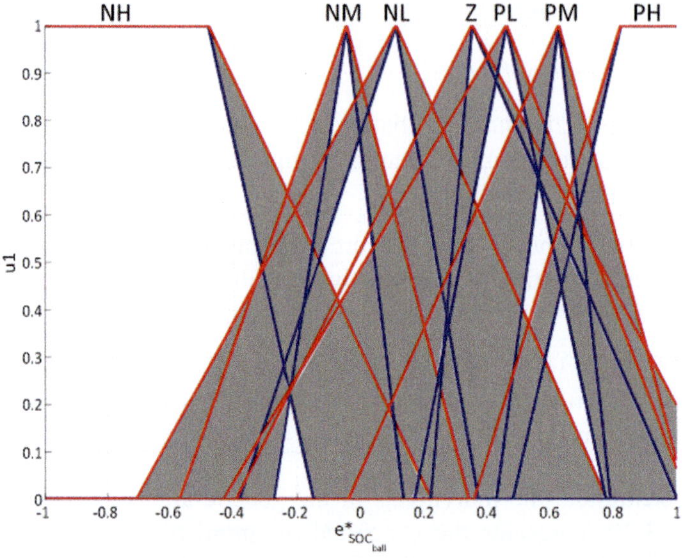

Figure 3.27. *Affiliation functions of input 2. For a color version of this figure, see www.iste.co.uk/jemei/cells.zip*

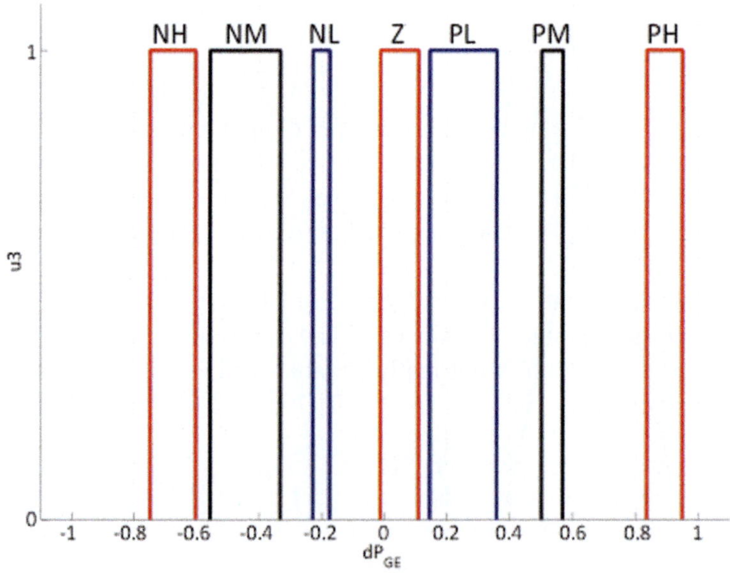

Figure 3.28. *Affiliation functions of the output. For a color version of this figure, see www.iste.co.uk/jemei/cells.zip*

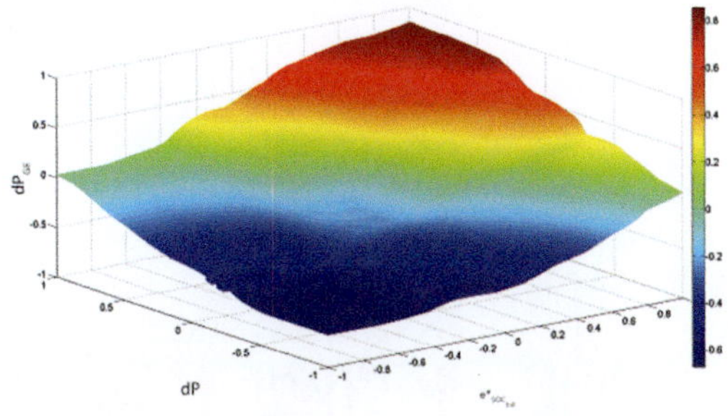

Figure 3.29. *Control surface. For a color version of this figure, see www.iste.co.uk/jemei/cells.zip*

3.4.4.2. *Results of the energy management strategy*

Figure 3.30 represents the evolution of various powers delivered, recovered by the energy sources or dissipated by the rheostat. Positive power corresponds to consumption, and negative power corresponds to recovery or dissipation. The mission profile is also represented. This figure shows a distribution of powers as implemented and imposed by equations [3.20] and [3.21]. Figure 3.31 is a zoom-in view of Figure 3.30.

It can be verified that, as indicated by the specifications, ultracapacitors are subjected to high frequencies of the mission profile, and the batteries, as well as the generator, are subjected to low frequencies. Similarly, batteries can be recharged due to the operation of the generator.

Figure 3.32 shows the juxtaposition of the evolution of the state of charge of the electrochemical accumulators, as well as the acceleration of the locomotive. As specified by the energy management strategy, the state of charge of this secondary source is properly controlled depending on the acceleration of the vehicle and ranges between 70% and 90%. Similarly, Figure 3.33 shows the juxtaposition of the evolution of the state of charge of the ultracapacitors, as well as the speed of the locomotive. The state of charge of the ultracapacitors is properly controlled depending on the speed of the vehicle and ranges between 50% and 100%.

The proper stability of the whole system can be noted and validated based on Figure 3.34. Indeed, the bus voltage (common voltage imposed to the set of elements) remains stable throughout the mission. A divergence of this voltage would lead to a full stop of the system, and even to its degradation.

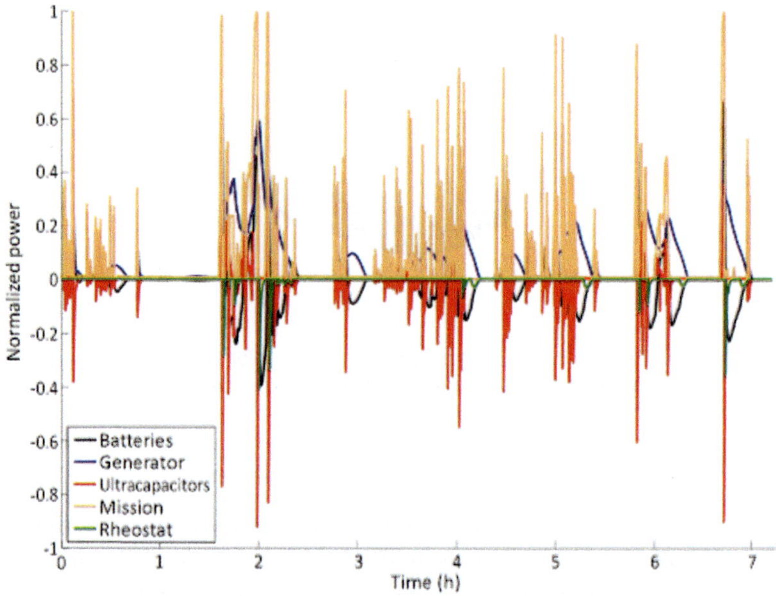

Figure 3.30. *Evolution of powers. For a color version of this figure, see www.iste.co.uk/jemei/cells.zip*

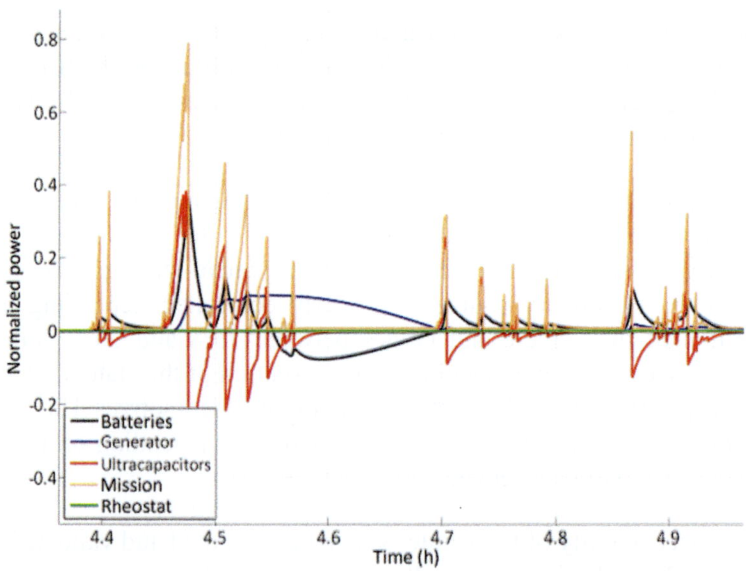

Figure 3.31. *Zoom-in view of the evolution of powers. For a color version of this figure, see www.iste.co.uk/jemei/cells.zip*

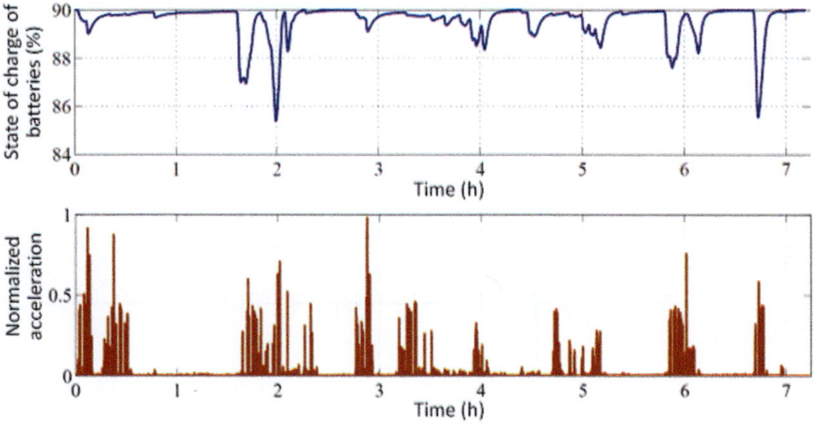

Figure 3.32. *SOC of EA and acceleration of HEL. For a color version of this figure, see www.iste.co.uk/jemei/cells.zip*

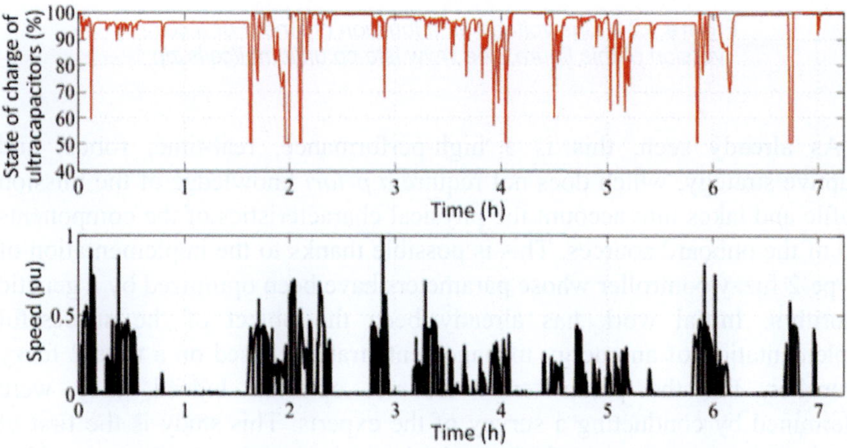

Figure 3.33. *SOC of UC and speed of HEL. For a color version of this figure, see www.iste.co.uk/jemei/cells.zip*

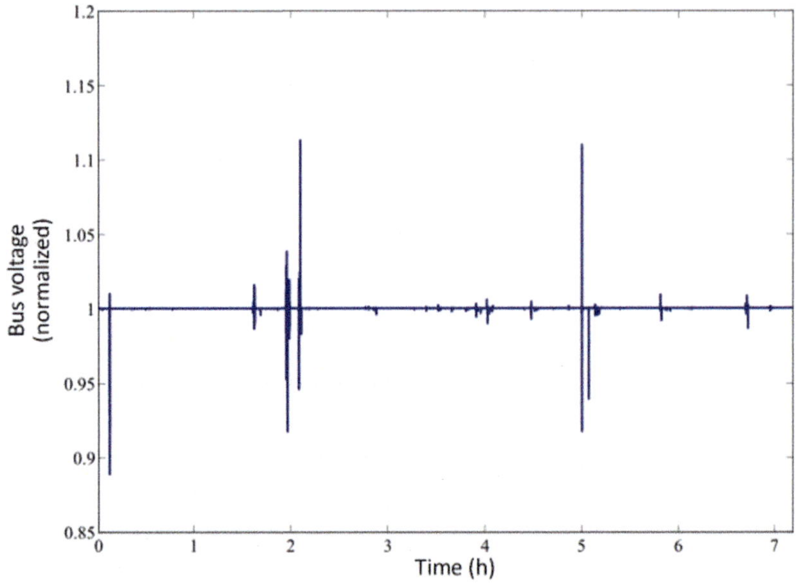

Figure 3.34. *Bus voltage as a function of time. For a color version of this figure, see www.iste.co.uk/jemei/cells.zip*

As already seen, this is a high-performance, real-time, robust and adaptive strategy, which does not require *a priori* knowledge of the mission profile and takes into account the physical characteristics of the components and of the onboard sources. This is possible thanks to the implementation of a type-2 fuzzy controller whose parameters have been optimized by a genetic algorithm. Initial work has already been the object of the successful implementation of an energy management strategy based on a type-2 fuzzy controller, but the parameters were not optimal. Indeed, they were determined by conducting a survey of the experts. This study is the first of this type to propose type-2 fuzzy logic optimized by genetic algorithm intended for hybrid electric vehicles. The perspectives of this study aim to improve the genetic algorithm. It would be interesting to include constraints and limitations in this optimization problem or other criteria to be minimized or maximized. For example, it is well known that the more the electrochemical accumulators are subjected to a large number of cycles, the more they are degraded. Minimizing this number of cycles would maximize the lifetime of the electrochemical accumulators and consequently minimize their replacement cost in time.

3.5. Conclusion

This study has allowed us to highlight two types of energy management for different applications. One of them relies on mathematical tools, and the other one on artificial intelligence tools.

In the first case, the wavelet transform, ARIMA models or the neural networks have enabled the development of an online energy management system applied to a heavy hybrid vehicle composed essentially of a fuel cell system, batteries and ultracapacitors. Thanks to the developed algorithms, and particularly to those relying on adaptive NARNNs, an energy distribution taking into account the frequency ranges of various sources has been implemented by using only the current and previous data of a univariate signal of the power demand of the vehicle. Thus, this online distribution between various sources makes it possible, among other applications, to increase the lifetime of the FC and batteries without *a priori* knowledge of the mission profile.

In the second case, energy management based on type-2 fuzzy logic and genetic algorithms has been proposed in order to minimize fuel consumption of a hybrid locomotive. In this case, the management takes into account the characteristics and dynamics of various sources of energy. Using the artificial intelligence tools, the following specifications have been met: limitation of the states of charge of the batteries, a distribution of energy demand consistent with the lifetime of the batteries and maintenance of the bus voltage.

Moreover, these works have shown that hybridization of energy sources makes it possible to obtain better performances (lifetime, reliability, efficiency, etc.). Nevertheless, the other sources of energy (EA and UC) are nowadays very reliable, and the efforts should focus on the lifetime of FCs. Indeed, hybridization seems to be necessary for transport and/or stationary applications. To obtain a high-performance system, it is important to have energy sources with the same degree of reliability. For this purpose, studies focusing on the diagnosis and prognosis of FC generators are essential. Chapter 4 is dedicated to these subjects.

4

Diagnostics and Prognostics of Fuel Cell Generators

4.1. Introduction

The previous chapters have highlighted the scientific and technological challenges stemming from the optimization and efficiency of the fuel cell system. It is nevertheless essential to take interest in a third challenge related to the lifetime of FCs. Indeed, wide-scale deployment of the FC will not take place unless it is sufficiently robust to be of continued service to the client. This is a major point to be studied upstream, together with the development of reliable indicators of their failures. The cost–performance–durability triad is essential for the competitiveness of the FC compared to other more conventional solutions.

Moreover, in order to transfer this technology to more efficient industrial applications, it is obviously essential to increase its lifetime, as well as to render the system more reliable. This involves an efficient control system to ensure the preservation of the FC high performances throughout its lifetime and the detection of possible faulty components. It is therefore necessary to develop diagnostic methods and tools. Diagnostics can be used in various situations:

– to identify the origin of failure, once it has occurred, in order to inform the user on its nature;

– to detect in real-time a deviation from normal operating conditions in order to inform the regulation and control system so that it corrects the deviation on its own by acting on its causes in order to predict its duration and impact on the performances and lifetime of the FC and/or of the system.

For the same purpose of significantly increasing the lifetime of the FC, it is important to understand its operation and control in order to slow down degradations before failure occurs. To this aim, the PHM discipline enables the continuous follow-up and estimation of the state of health of a system, the prediction of its remaining useful life and decision-making for its preservation, so that it successfully completes its mission. The prognostics of FC systems operation is a field worthy of further exploration aimed to implement counter-measures for increasing the remaining useful life of the FC system and also to define realistic guarantee periods that meet the market expectations.

There are many approaches to the development of diagnostic and prognostic tools. However, it is important to have prior in-depth knowledge of the degradation mechanisms of the FC and of its system in order to find efficient solutions.

The first part will explore the degradation mechanisms and the failures that may occur on the FC and its system. The second part will describe the diagnostic methods applied to FC systems. Finally, the third part will describe the prognostic tools.

4.2. Degradation phenomena in a fuel cell and its system

Degradation is defined as the act of damaging, altering or deteriorating an object. A degradation process can lead to failure in a system. These degradations are generally caused by a process of alteration generated by changes in the inherent structural properties of a system, irreversibly affecting its current and future performances [JOU 15a].

In a FC system, the central element, and the most sensitive one to degradations and/or failures, remains the FC stack. Nevertheless, lifetime- and reliability-related problems also concern the actuators that ensure the proper operation of the FC. The full system is therefore worthy of attention in our work.

It is important to be able to analyze and classify the degradation mechanisms that can manifest in a fuel cell system. Several classifications can be advanced, depending on the study object. Indeed, degradations can be studied by family (chemical, thermal, mechanical, etc.), by individual component constituting the stack (membranes, electrodes, GDL, etc.) or by failures.

Thus, various combinations of factors can influence the reliability of these complex systems, some of which play a major role in the activation and acceleration of degradation, while others have a minor influence. The extent and severity of the failure depend on the nature, size and duration of exposure to these factors.

The following sections offer a classification of the various categories of failure that can occur in a fuel cell system.

4.2.1. *Reversible or irreversible degradations?*

The set of degradation mechanisms occurring in a FC system can be classified depending on their degree of severity:

– Reversible degradations affect the performance stability of the FC system. A corrective action on the operating conditions proves to be sufficient for the system to regain its initial performance. This can happen in various situations (faulty control, experimenter's error, punctual failure, etc.).

– Irreversible degradations occur when it is no longer possible for the FC system to regain its initial performance even after the cause of the failure has been eliminated. Besides affecting the performance, this type of degradation results in permanent physical and structural modification of the materials.

4.2.2. *Degradations of FC components*

As the first step in describing the degradations that can occur in the auxiliaries of a FC system, it is important to present the degradations inherent to the components of the FC, since certain degradations of the FC are directly linked to the failure(s) of the actuators surrounding the FC.

Let us therefore quickly review the degradations of FC components. More detailed studies can be found in [JOU 15a, KOC 12, KUN 06, SAN 15, SCH 08, YOU 09c, TAW 12, WAS 10, WU 08a].

It is worth recalling that the FC stack contains a stack of unit cells. The failure of a single cell can trigger the failure of the full FC system. There are obviously various causes leading to cell degradation. A list of cell components and their corresponding degradations is provided below [JOU 15a]:

– Degradation of bipolar plates:

- corrosion affects the durability of the membrane and the catalytic layers;

- emergence of a resistive surface layer leading to an increase in ohmic resistance;

- deformation or crack of the plates due to too high current densities or inadequate thermal cycles.

– Degradation of gas diffusion layers (GDL):

- loss of hydrophobicity;

- change of the structure of diffusion layers due to carbon corrosion and to mechanical constraints;

- loss of porosity.

– Degradation of sealing gaskets following mechanical or thermal disturbances leading notably to leakages.

– Degradation of electrodes:

- Loss of active surface of the electrode due to degradation of the catalytic layer or the carbon support:

- degradation of the manifold when the FC operates at high voltages close to OCV;

- the process of carbon corrosion is accelerated when the FC is subjected to stop/start operations;

- degradations due to current dynamics.

- The manifold is sensitive to poisoning with CO, SO_2, H_2S, NO_2, NO and NH_3, which occupy the active sites and prevent reactions from taking place.

– Degradation of the membrane:

- chemical degradation caused by the contaminants that degrade the polymer;

- mechanical degradation that can lead to membrane tearing;

- short circuit: electrons cross the membrane.

These degradations can also be grouped according to their mechanical, chemical or thermal origin.

4.2.3. *Degradations depending on origin*

4.2.3.1. *Mechanical origin*

Mechanical degradations occur especially at the membrane level. Improper fabrication process, the presence of foreign particles or inadequate operating conditions can lead to punctures, tears or pinholes in the membrane. These failures are often considered the main cause of premature failures [X12].

Nevertheless, mechanical degradations can also appear as a result of improper hydric or thermal management of the FC. Indeed, an increase in relative humidity can generate a swelling pressure in the membrane, which can lead to the propagation of microcracks [MEN 11, WAN 11a]. It is worth noting that among the mechanical processes that generate a significant degradation in the PEMFC, humidification or drying cycles play an essential role in generating membrane dilation or contraction. This leads to mechanical strain of the membrane and of the seal gaskets.

4.2.3.2. *Chemical origin*

Chemical degradations are essentially due to the contaminants in the ambient air or in the fuel. Atmospheric pollutants include NO_x, SO_x, CO and CO_2. They directly involve in the process of corrosion of the components (bipolar plates and carbon support in the catalyst) and generally drive a reversible degradation.

On the contrary, fuel contamination by certain atmospheric contaminants such as H_2S and SO_x can cause irreversible losses of performance. This contamination affects the kinetics of the electrodes, conductivity and mass transfers, which are essential for the performance of the FC [COL 06, KEL 05].

4.2.3.3. *Thermal origin*

When the PEMFC operates at temperatures above the optimal temperature, its membrane may get dry, rendering it more electrically resistive, while when it operates at temperatures below the optimal temperature, water may accumulate in the cells, leading to a decline in performance due to an increase in the effects of mass transport (flooding).

It is also possible to have local hot points due to various phenomena (hydrogen depletion, short circuit, nitrogen accumulation, etc.). Local overheating may cause a tear of the membrane.

4.2.4. *Failures of the FC system*

In the previous sections, we have presented the classification of the various degradations that may occur in a FC stack. In a FC system, the stack is the central element, and failure of one of the surrounding actuators leads to a degradation of the performances of the full system. Figure 4.1 shows the location of the failures of a FC system. The possible consequences due to these failures are listed below [SAN 15].

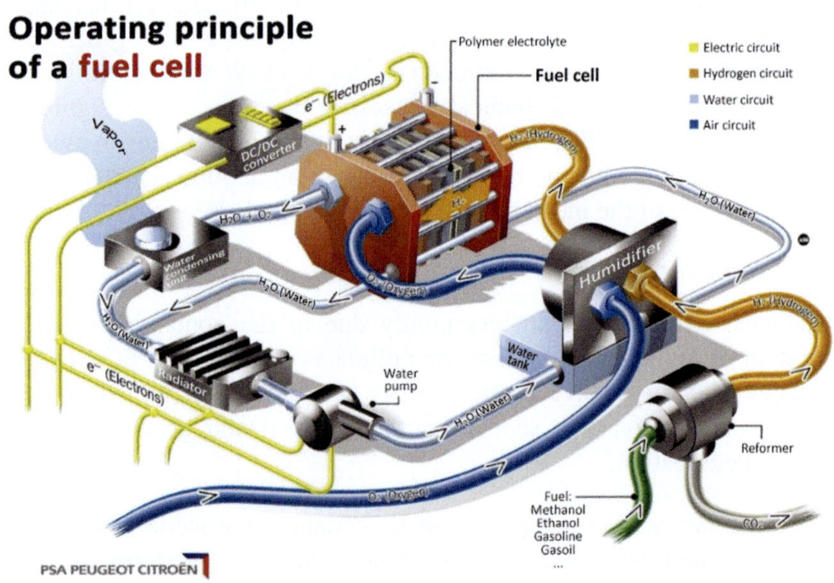

Figure 4.1. *Locations of failures of a FC system. For a color version of this figure, see www.iste.co.uk/jemei/cells.zip*

4.2.4.1. *Anode and cathode circuits*

The failures related to the anode and cathode circuits often concern the fuel and oxidizer supply of the cathode and anode compartments. If a compressor unit (MCG) is used for supplying the cathode with air, then malfunctions related to the speed of the MCG may occur. In this case, the FC is undersupplied (air starvation). Control-command problems, improper generation of the air or hydrogen flow rate set point and a delay in the generation of this set point and/or of the real value of the flow rate can also cause improper supply. In a PEMFC, when there is reactant undersupply of the cells, their individual voltages diminish compared to the others, and there is increased scattering of cell voltages. During a "weak" undersupply with reactants, the voltages of undersupplied cells remain positive, though lower than the others. On the contrary, when the flows of reactants are not sufficient to maintain the required level of current, certain cells have their voltage drop to negative values, due to their operation in "electrolysis" mode. Cell reversal is the cause of manifold dissolution and agglomeration, as well as of the carbon corrosion, leading to a loss of catalytic surface [LIU 06, YOU 09c, TAN 08].

Gas flow rates also have a significant influence on fuel cell humidity. They play a significant role in water evacuation within channels and diffusion layers. When water is in excess, water vapor in the cells may condense and block gas transportation in the electrodes, diffusion layers and gas channels [YOU 09c]. Losses by mass transport occur when there is a gap between the rate of yield of reactants and the speed with which they are consumed. This disparity can be generated by a low or high stoichiometric factor.

In the first case, the distribution of reactant gases in the FC cells is not sufficiently homogeneous for each cell of the fuel cell. Liquid water droplets may clog the flow channels or the diffusion layers, which may lead to variations of cell voltages.

In the second case, a high stoichiometric factor in air coupled with insufficiently humidified reactant gases can dry the fuel cell stack and drive an increase in ohmic losses, limiting the migration of protons inside the electrolytic membrane. A high air flow rate facilitates a better distribution of reactant gases in the distribution channels, therefore inducing an increase in the oxygen concentration in the catalytic layer, which leads to improved FC performance. It should be noted that a significant and abrupt variation of the flow rate may lead to the tear of the electrolytic membrane due to pressure variation in the cathode compartment.

4.2.4.2. *Coolant circuit*

The coolant circuit is essential for the PEMFC operation. A decrease in the cooling water flow rate may lead to an uneven distribution of flow through the channels, which generates local overheat. A short-term increase in FC temperature may lead to erosion of the active surface and generation of hot points. Moreover, excessive heat generation may cause the tear of the membrane and even its lighting [KUN 12, LAG 13, MAT 13].

At high temperature, it is possible that membrane materials in the catalytic layer are not fully hydrated. Membrane conductivity and the active surface of the catalyzer diminish due to lower relative humidity of the reaction gases and the reduction in the water content of the membrane. When the temperature increases, the water evaporation rate becomes higher. When temperature reaches a critical value or the amount of water evaporated exceeds the amount of water produced, the membrane starts to dry.

The causes of these malfunctions include a decrease in the speed of the coolant circuit pump, a problem with the primary or secondary exchangers or a control/command problem.

4.2.4.3. *Electric circuit*

The fuel cell is a high-current, low-voltage generator. Specific power converters should therefore be developed for voltage regulation according to load requirements.

Indeed, voltage depends on the required current, as well as on temperature, partial pressures of the reactive gases, hygrometry, stoichiometric factors and fuel cell ageing state. A FC can be subjected to strong constraints during its use in transport applications: unless hybridized, it always operates in strongly variable loading state, the required power following that of the electric traction chain. If hybridization exists, the power demand at the fuel cell level is not directly correlated to that of the traction chain, but is still highly variable.

The failures of the converter can be related to control-command, as well as to the failure of a physical component of the converter. Indeed, many works [BRY 11, CHA 10, KUL 10, WAN 13a, WAN 13b, WAN 14, YAO 13] show that electrolytic capacitors and semiconductor devices of the converter get more rapidly degraded than other components. The degradation of these components has a significant influence on the operational efficiency of the DC/DC converters: it causes the voltage at the converter output to decrease to a value that is below specifications, eventually leading to the breakdown or failure of the whole system. Short circuit is the most unfavorable situation, during which excessive increase in temperature in the fuel cell stack leads in several seconds to irreversible degradations [SAN 14b].

4.2.4.4. *Control circuit*

An important factor to consider in the control dynamics is the diversity of time constants applying to the PEMFC (different fluidic, thermal and electrochemical time constants). The response time of the fuel cell is limited by the constraints on the gas flow rate, anode and cathode pressure, temperature and hydric pressure. The supervisor needs to permanently make sure that the operating conditions of these parameters are optimal. The system instrumentation and control must be reliable and robust (auxiliaries generate many disturbances during transient and failure phases) and need to adapt to the state of degradation of the fuel cell stack. Moreover, it is preferable to use non-invasive sensors in order to limit the disturbances in the FC stack, especially at the level of fluid flows.

Hence, an evolving instrumentation and control requires integrated non-invasive diagnostic methods.

Figure 4.2 presents a synthesis of operating factors influencing reliability in a PEMFC [KUN 06]. It is worth noting that even if the degradations considered have various causes and involve various elements, the consequences are always loss of power, instability and decrease in the lifetime of the fuel cell.

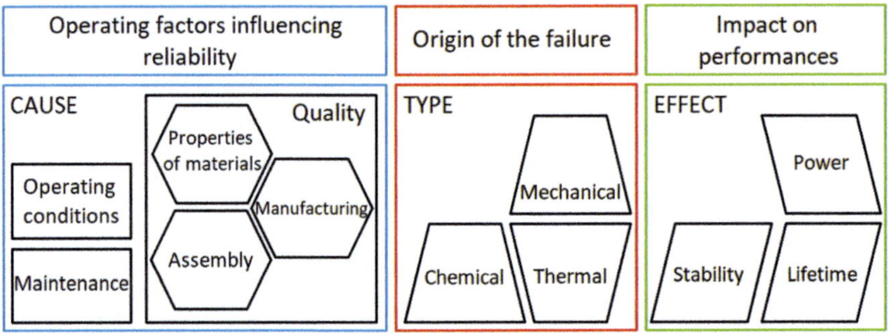

Figure 4.2. *Representation of failure modes in a PEMFC. For a color version of this figure, see www.iste.co.uk/jemei/cells.zip*

4.3. Diagnostics

The previous sections have enabled us to provide a non-exhaustive list of the degradations and failures that can affect the stack of a FC or a FC system. The main purpose of diagnostics is to identify the possible cause of the failure(s) using logical reasoning based on a set of information resulting from an inspection, control or maintenance test [AFN 88].

In general, the appearance of a system equipment fault can generate failure, in the sense of loss of a function of the system, which is no longer operating normally. In order to determine whether a system properly meets its objectives, its operation must be precisely monitored by means of sensors strategically positioned or indicators. This monitoring enables us to detect the system's passage to abnormal operation [RIB 09]. For a FC system, there are many measurable quantities (flow rates, pressures, temperatures, hygrometries, individual cell voltages, current as well as internal parameters of the FC), and it is obvious that the use of all these sensors would make it possible to establish an efficient diagnostics. Nevertheless, our aim is to provide solutions that are technically and economically viable, and to avoid as much as possible the introduction of intrusive sensors that may disrupt the operation of the FC. Thus, [HIS 06] proposed a classification of measurable quantities on a FC generator (Table 4.1). Throughout the works presented in this book for the development of diagnostic tools, the measurement of the FC voltage or the measurements of cell voltages or the results obtained during dynamic characterizations (impedance spectroscopy) of the stack will be used.

Measurements that are technically or economically possible	*Measurements that are technically or economically difficult*	*Technically or economically impossible measurements*
– FC current – FC voltage – FC cooling water temperature – H_2/air temperatures (inputs and outputs)	– Cell voltages – H_2/air pressures (inputs/outputs) – Internal temperatures	– H_2/air flow rates – Air/H_2/water channel flow rates – H_2/air hygrometry – Water content of the membrane – Stack impedance per EIS – Quality of input gases and effluents by mass spectrometry

Table 4.1. *Quantities measurable on a FC generator [HIS 06]*

The focus is on the data issued from the FC, which can itself be seen as a sensor. Indeed, a defect occurring on one (or more) auxiliary component(s) has repercussions for the operation of the FC stack. Hence, degradations, errors or failures can be detected by means of stack [CAD 14].

In this context, for many years, our work has focused on research related to the subject of diagnostics. These activities rely particularly on the ANR DIAPASON2 project and the thesis by Elodie Pahon [PAH 15b]. The main objective of the thesis has been to capitalize on the approaches of FC diagnostics that our team has considered since 2005. Within the same context, the ANR DIAPASON2 project has opened the path to realizing onboard diagnostic systems.

Thus, the first part proposes an overview of diagnostic methods applied to the FC and the second part presents the methods developed in the thesis by Elodie Pahon [PAH 15b] and the ANR DIAPASON2 project. The last part is dedicated to the results obtained with these methods.

4.3.1. *Diagnostic methods applicable to fuel cells*

The aim of this section is not a state of the art, but rather a presentation of the many diagnostic methods applicable to FCs. They rely especially on knowledge of the system and the data or signals issued by the FC system. For these techniques, it is important to conduct many experimental tests for obtaining references on the normal operation of the FC (rated behavior) and/or on abnormal operation (behavior in the presence of errors). This reference is represented by a history, an experiment or a known or estimated model of the system's behavior [RIB 09]. A classification of these methods is proposed in Figure 4.3. Knowledge-based methods are only developed (rule-based systems, FMECA, fault trees, etc.), as they are hardly ever used for the FCs.

Thus, two paths have been developed for the diagnostics of fuel cells: one based on a model and the other based on shape recognition. The first one compares the values of measured variables to those predicted by a model that is physical, behavioral (black box or hybrid (gray box). The physical models explicitly take into account the phenomena governing the system, simplifying the diagnostic process, but they require knowledge of the values of internal parameters, which are difficult to measure and/or require dedicated sensors. These sensors render the stack and the system more complex, increasing its cost and reducing its reliability. As for the black box models, they require neither knowledge of the physical relations governing the processes to be modeled nor an identification of the internal parameters of the fuel cell. However, the absence of explicit causality renders failure localization more complicated. Signal-based methods require no model and rely only on *a priori* knowledge and experience feedback. They only involve signal analysis (voltage, current, etc.).

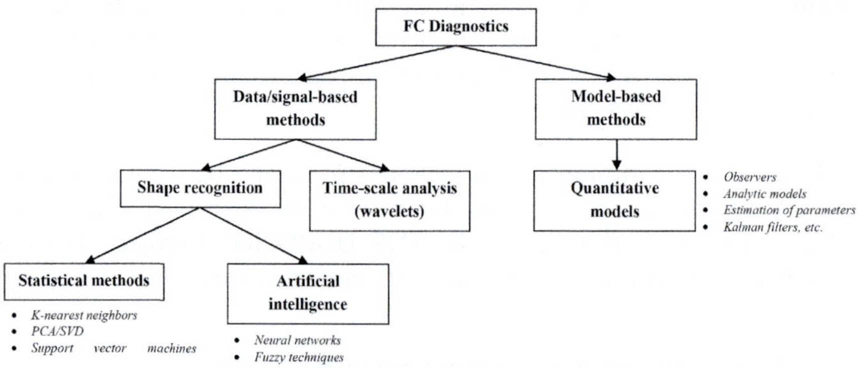

Figure 4.3. *Classification of diagnostic methods*

4.3.1.1. *Model-based approaches*

The development of mathematical models involves the use of a series of algebraic and/or differential equations. They require internal system parameters, which are sometimes difficult to identify, particularly so for fuel cells (size of channels, humidity rate of the membrane, etc.). In-depth physical knowledge of the system is therefore required when developing model-based diagnostic approaches. Moreover, the calculation times for the development of this approach are often very long, which renders online diagnostics even less probable.

Nevertheless, the mathematical model (which is also called the knowledge-based model or the "white box" model) makes it possible to very precisely describe the studied system under normal conditions and even when failure occurs. The diagnostic method then relies on the comparison between the real observed behavior of the physical system and the model-based predicted behavior. Incoherence detection helps in drawing a conclusion on failure occurrence in the system. A malfunction model (failure model) makes it possible to locate and identify likely failures [BRE 15, CHA 13, RIB 09, SOR 09].

A state of the art of model-based diagnostic methodologies applied to PEMFC was proposed by [PET 13]. A similar study was conducted in [WAN 11b], but it is focused on SOFC. This will not be elaborated here, as it exclusively focuses on data-based approaches.

4.3.1.2. *Data-based approaches*

Data-based approaches involve the use of "black box" models for classification. They identify the normal behavior of the system to be diagnosed, and their main advantage is that they do not require in-depth physical knowledge of the process. Moreover, they are often less expensive than the approaches based on knowledge models, which renders them quite compatible with an online application.

These approaches rely essentially on an available set of experimental observations on the system. The aim is then to associate a set of measurements conducted on the system with known operating states.

These data-based approaches often call for shape recognition methods that use numerical training and classification techniques in order to establish a reference model of the system based on experience (data processing, history of measurements). This reference model captures the normal behavior of the system and is used for detection and diagnostics [RIB 09].

A state of the art of data-based diagnostic methodologies applied to PEMFC is proposed in [ZHE 13a].

Diagnostics based on shape recognition (SR):

Shape recognition (SR) relies on the classification of objects or forms by comparing them to typical shapes. The present work uses uniquely statistical SR, which is based on the numerical representation of shapes [DUB 90] (structural SR uses shape grammar representation [FU 74]).

The purpose of diagnostics by (statistical) shape recognition is to assign a class to an object, the latter being described by a set of parameters called "descriptors", "variables", "attributes", "parameters", "characteristics" or "shapes". In a monitored context, K classes (Ω_1, … Ω_K), in which observations x must be classified, are known in advance. Vectors belonging to the same class form a cloud of points occupying an area in space. In order to classify a new observation, the SR-based diagnostic strategy delimits the areas, determining a separation boundary between various classes, thanks to a set of known observations, called training set. This stage enables the characterization of classes and the definition of decision boundaries. Once the classes are defined, the algorithm can be used to classify new observations [YOU 09c].

The main classification techniques used to build such a model can rely on statistical methods (k-nearest neighbors, support vector machine, etc.) or on artificial intelligence (neural networks, fuzzy logic, etc.). Many works use shape recognition-based diagnostics for fuel cells [HIS 07, KIS 10, LI 14a, LI 14b, ONA 12, PAH 14a, PAH 14b, PAH 15a, PAH 15b, PAH 16, WAS 10, WU 08b, YUA 07].

Diagnostics based on timescale methods:

Diagnostics based on time-frequency methods has been used for many years in certain domains, and notably for rotating machines and internal combustion engines [WU 05]. The Fourier transform enables a frequency content to emerge from a time signal. This content is easily identifiable under healthy operating conditions, the occurrence of failure leading to a deviation of the frequency content of the measured signal. The amplitude and position of the harmonics are finally a signature of the state of the process to be diagnosed. Fourier transform is unfortunately not efficient for stationary signals whose statistical properties are invariant (or vary very little) in time. In the case of FCs, the signals involved are non-stationary.

Fourier transforms with sliding window could be adapted to measure the time variations of the frequency components of a signal. Nevertheless, the time-frequency resolution of the wavelet transform is more flexible and more optimized [MAL 99]. The wavelet transform enables the adjustment of the size of the window according with the position in the time-frequency plane. The components of the same signal do not necessarily evolve in the same manner, and their behavior varies depending on their low or high frequency range. A low-frequency component requires an observation time that is sufficiently long to be correctly analyzed. On the contrary, a high-frequency component rapidly evolves and requires a shorter time to be analyzed. It is precisely the adaptability of the analysis of the wavelet transform that enables us to analyze these components in a different manner [HAM 08].

The use of this mathematical tool for FC diagnostics has proved its efficiency for highlighting various failures (flooding, dry out, oxidizer undersupply, excessive temperature, etc.) [BEN 14a, PAH 15b, STE 11, WAN 12, YOU 09b, ZHE 13b].

Further methods have been developed for FC diagnostics. They rely on the definition of reference signatures representing known normal and faulty operating modes of the system. Among these signatures, the acoustic signatures [TAN 13] and magnetic signatures [HAM 14] are worth being mentioned. Diagnostics involves a comparison of current signature with predefined reference signatures.

4.3.2. *Diagnostic methods developed*

It should be recalled that regardless of the data-based diagnostic method considered, a significant amount of data on the real system under normal operating conditions and also when failures occur should be collected. Hence, various FCs have been subjected to several thousand hours of testing within FCLAB Federation in the context of the theses of Rosa Elvira Sanchez [SAN 15], Elodie Pahon [PAH 15b] and Simon Morando [MOR 15a].

4.3.2.1. *Diagnostics based on shape recognition coupled with k-NN*

For this type of method, the general procedure that should be followed is presented in Figure 4.4 [YOU 09c]. It involves the following:

1) setting up a database acquired on the system;

2) building a representation space: the purpose of this stage is to optimize the representation space. Indeed, there can be many parameters measured on the system, but not all of them are significant. The number of characteristics should therefore be diminished, while maintaining high discrimination capacity. This stage offers several advantages, such as the reduction of calculation time, simplified classification, method robustness and reduction of the number of sensors *a posteriori* present on the system. There are various tools that enable the optimization of this space (linear or nonlinear analysis of the main components, linear or nonlinear analysis of Fisher discriminant, etc.) [LI 14a, PAH 15a, YOU 09c]. They are used for obtaining the descriptors;

3) training and defining the decision boundaries: training is required for supervised classification. Indeed, descriptors are associated with a (healthy or faulty) state of the system. Several classes are then established and labeled. Then, decision boundaries are defined between classes. Consequently, at each new measurement point (or new observation of the system), a decision should be made on the classification manner. It is the classification stage.

There are several rules for detection and classification, among which, the following methods are worth mentioning: k-nearest neighbors (k-NN) [PAH 15a], support vector machines (SVM) [LI 14b], Gaussian mixture model (GMM) [WAS 10] or methods based on artificial intelligence (neural networks, fuzzy logic [ZHE 14], etc.).

The following section presents the k-NN method, which has essentially been studied in the works of Elodie Pahon.

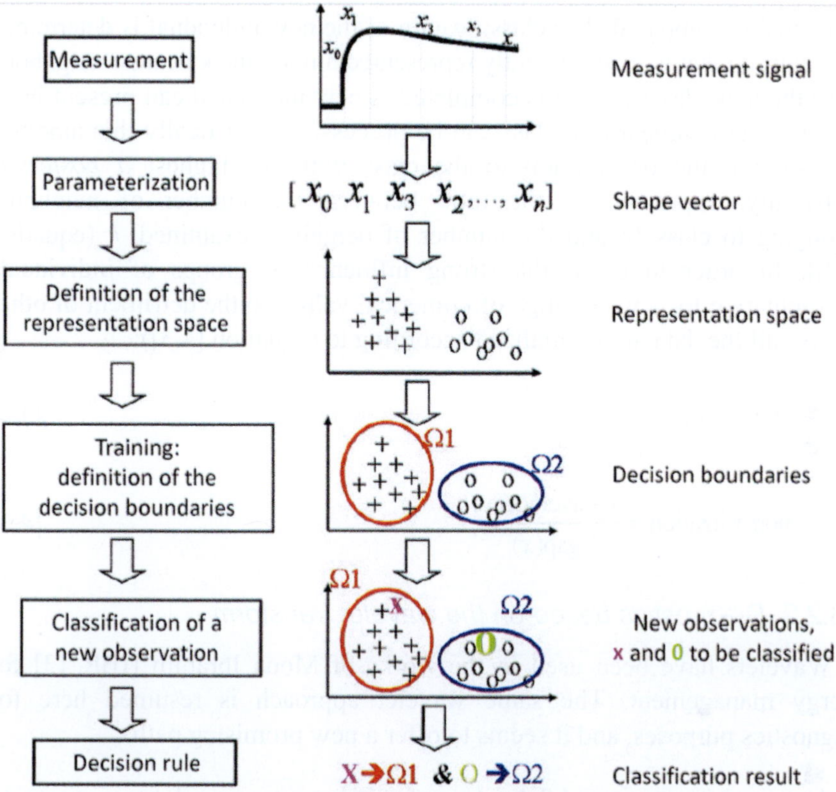

Figure 4.4. *Shape recognition-based diagnostic procedure. For a color version of this figure, see www.iste.co.uk/jemei/cells.zip*

The k-NN rule examines the distances between a new observation (or individual) to be classified and already classified samples (according to a predefined metrics) during the training phase and then assigns the new observation to the most strongly represented class among its k nearest neighbors [RAU 01, YOU 09c]. For this purpose, the new individual is compared to all the individuals in terms of distance. Due to its simplicity and good results during classification, Euclidian distance (equation [4.1]) is used:

$$\text{dist}(x, y) = \sqrt{\sum_{i=1}^{n} (x_i - y_i)^2} \quad [4.1]$$

Finally, it is important to define the surrounding environment that allows for the choice of a value of *k* number of neighbors to which the new individual is compared. The classification of the new individual is determined by the group that is predominantly represented among the k nearest neighbors. Once the new classification is completed, a new individual can present itself so that its assignment to a class can be processed. Statistically this amounts to assigning the observation to the class with the highest *a posteriori* probability, expressed as the ratio between the number of neighbors belonging to class Ω_j and the number of neighbors examined, k (equation [4.2]). In order to avoid the strong influence of groups of individuals representative for a wide range of numerical values to the detriment of other groups, all the data are normalized according to (equation [4.3]):

$$P\left(\Omega_j / x\right) = \frac{k_i}{k} \qquad [4.2]$$

$$\text{normalization} = \frac{x - \text{average}(x)}{\text{gap}(x)} \qquad [4.3]$$

4.3.2.2. *Diagnostics based on the wavelet transform*

Wavelets have been used in the works of Mona Ibrahim [IBR 13] for energy management. The same wavelet approach is resumed here for diagnostics purposes, and it seems to offer a new promising path.

As seen in section 4.3.1, decomposition into wavelets amounts to successive filtering of the signal using a high-pass filter and a low-pass filter. The use of this band pass filter makes it possible to obtain an approximation signal (low frequencies) and a detail signal (high frequencies). Let us also recall that the detail signal D_n contains high-frequency information within the range $\left[\frac{f_s}{2^{n+1}}, \frac{f_s}{2^n}\right]$ while the approximation signal A_n contains low-frequency components of the signal *s(t)* within the range $\left[0, \frac{f_s}{2^{n+1}}\right]$, where n is the decomposition level. An example of wavelet decomposition of a signal *s(t)* is given in Figure 4.5. In order to recompose the signal, it is sufficient to make the sum of the set of detail and approximation signals.

The choice of wavelet or of the decomposition level will not be reviewed here, as they have been explained in section 4.3. Indeed, the focus will instead be on an original approach based on wavelet energies for the diagnostics. This method has been developed in the works of Kun Wang's thesis [WAN 12], a PhD student in our research team. Her works have enabled the diagnostics of a SOFC. The objective is to determine whether the energy contents of a defined signal evolve as a function of operating conditions and in what manner.

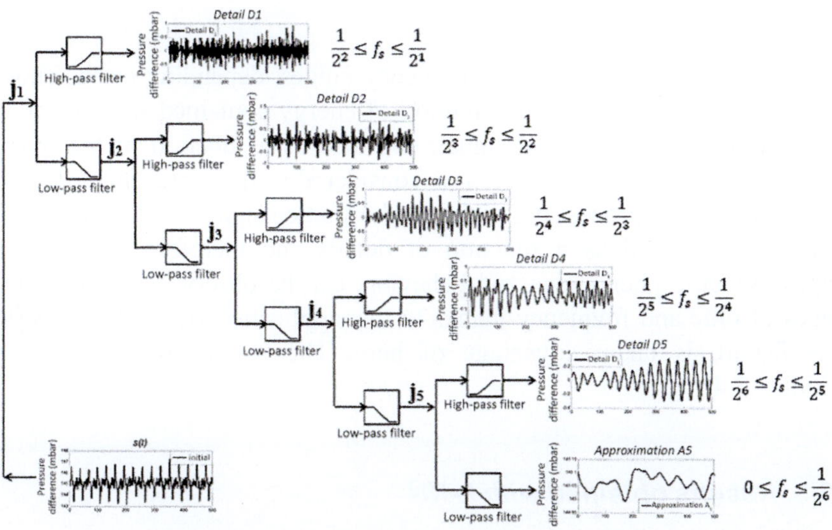

Figure 4.5. *Example of wavelet decomposition of a signal s(t) for five decomposition levels*

Given that all information transmissions involve energy transmission, relative energy contained in the decomposition signal would seem an excellent indicator for the change of state of the FC. Detail signals (D_n) are those containing most information. Therefore, the focus will be on the energy contained in the latter. The energy of a detail signal is defined by (equation [4.4]), while the total energy of signal *s(t)* is equal to the sum of energies of each of the detail signals (equation [4.5]):

$$E_n = \sum_{n=1} \left| D_n(k) \right|^2 \text{ where } k \text{ is the amplitude of each signal} \qquad [4.4]$$

$$E_{\text{Total}} = \sum_n \sum_k \left| D_n(k) \right|^2 = \sum_n E_n \qquad [4.5]$$

Finally, relative wavelet energy (RWE) is defined as the ratio between the energy contained in a detail signal and the overall energy of signal *s(t)* (equation [4.6]):

$$\text{RWE} = \frac{E_n}{E_{\text{Total}}} \qquad [4.6]$$

The energy is determined by frequency ranges, depending on the chosen decomposition rank *n*. The distribution of energy contained in a signal *s(t)* can be represented by detail signal, and also as a function of the overall energy via RWE. This distribution is considered as the representation of the information of the process described by the studied signal *s(t)* [PEN 14]. The energy is evaluated as a function of details and therefore of frequency ranges. Specific phenomena and behaviors can be observed within various ranges of time and frequency thanks to this application [ROS 01, ROS 04]. This feature is taken advantage of here. The results are presented in section 4.3.4.

4.3.3. *Results obtained with k-NN*

4.3.3.1. *Database setup*

As already noted, the first stage involves setting up a training database. Three databases linked to the use of three FCs of different sizes are presented. The failures tested for each of them are summarized in Table 4.2. Each failure generates a specific signature that can be identified by electrochemical impedance spectra (EIS). Indeed, depending on operation modes, their basic structures are preserved with two arcs (high and low frequency), but their shapes are modified. This change of profiles is therefore an indicator of the state of health of the fuel cell system [KUR 08, RUB 07, WAG 04, YUA 07]. Electrochemical impedance spectrum (EIS) is used as an indicator for failure diagnostics. Table 4.2 summarizes the number of EIS obtained depending on the tested FC and on the failure to be identified.

		Types of PEMFC		
		8 cells	***20 cells***	***40 cells***
Operating conditions	***Healthy mode – Rated***	60 spectra	20 spectra	100 spectra
	Flooding		80 spectra	
	Dry out		100 spectra	
	Air supply failure	55 spectra		210 spectra
	Short circuit			80 spectra
	Problem on the coolant circuit	12 spectra		60 spectra
	CO poisoning	19 spectra		

Table 4.2. *Composition of training bases*

Figures 4.6–4.9 illustrate two failures occurred on a PEMFC of 40 cells. The first involves diminishing the air supply at fuel cell input in order to simulate air starvation that may occur as a result of failure of the compressor unit or of its control. Figure 4.6 shows that the cathode stoichiometric factor (CSF) has been brought from 2 (rated point) to 1.6 and then to 1.4 (least favorable point). The signature of these failures is visible on the EIS obtained when various values of the CSF have been applied to the cathode of the FC. Indeed, Figure 4.7 shows that the stronger the undersupply, the higher the low-frequency arc of the spectrum.

The second failure to be presented here (Figure 4.8) involves increasing the internal temperature of the FC by diminishing the flow rate of the cooling water circuit. This failure can occur if the water recirculating pump or its control is faulty. Therefore, as a first step, this flow rate is halved in order to observe its influence on the temperature and on the voltage; then, water circulation is fully cut, which triggers an abrupt increase in temperature ($\Delta \approx 20°C$). Higher hot points have certainly appeared on the catalytic sites of the FC, but intrusive micro-thermocouples are needed to quantify them. Nevertheless, a very clear voltage drop is noted due to the dry out of the membrane (Figure 4.8). The EIS represented in Figure 4.9 have been drawn before and after failure occurrence.

Further failures have been tested on various FCs. It is worth noting that the failures mentioned in Table 4.2 are all reversible provided they are under control. Indeed, some of them can be considered irreversible if their application time is too significant (short circuit, cooling circuit and CO). In our case, well-defined protocols have been established in order to obtain the same performance as that measured before the failure occurrence.

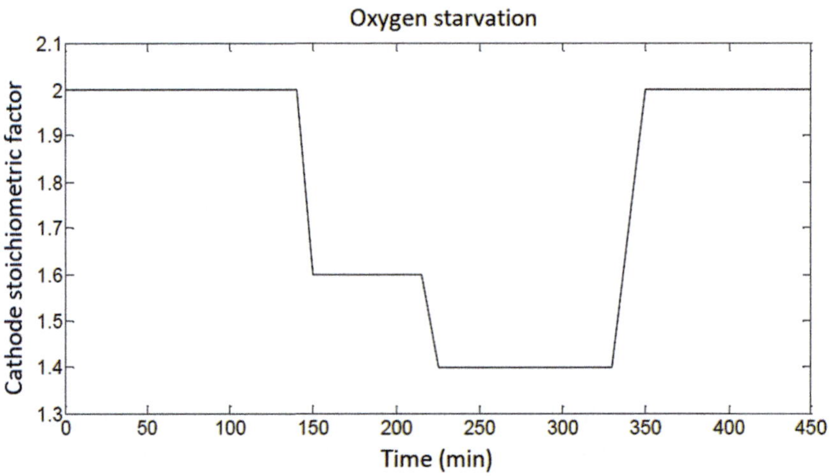

Figure 4.6. *Air starvation profile*

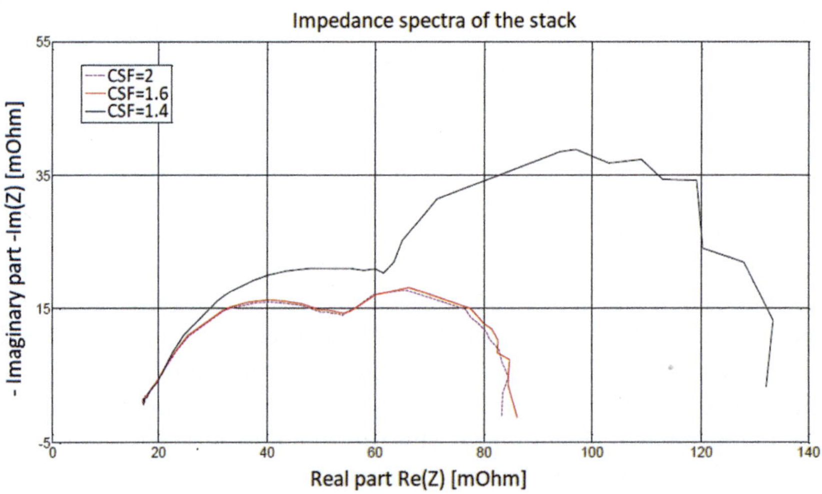

Figure 4.7. *EIS of a stack of 40 cells (air starvation). For a color version of this figure, see www.iste.co.uk/jemei/cells.zip*

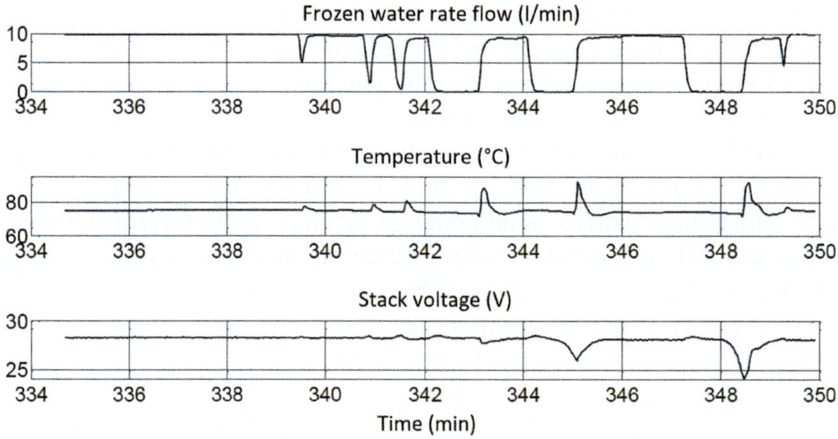

Figure 4.8. *Profile of water flow rate and its influence on temperature and on stack voltage*

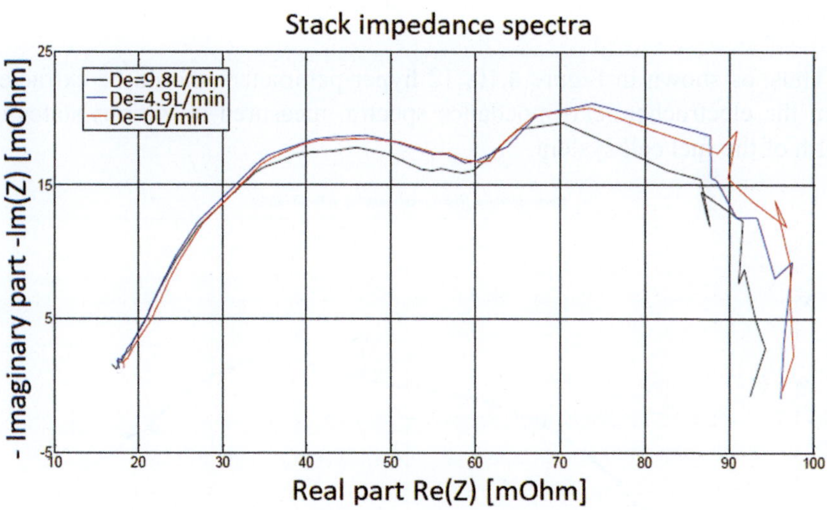

Figure 4.9. *EIS of a stack of 40 cells (temperature increase). For a color version of this figure, see www.iste.co.uk/jemei/cells.zip*

4.3.3.2. *Choice of descriptors and establishment of training base*

The experimental results presented in the previous section and summarized in Table 4.2 enable us to have substantial training bases for the classification. Nevertheless, let us keep the context of real-time diagnostics, which involves optimal use of the embedded memory space. The labeling phase of the k-nearest neighbor algorithm and an initial data training can be conducted offline. Nevertheless, the memory space should be available on the embedded system in order to store the database needed for testing phase, and for the online diagnostics. In order to optimize the occupation of this space, let us only extract several points of the impedance spectra and not the full spectrum. By relying on simple observations and more expert knowledge such as that which can be found in [ONA 12, WAS 10], we observe that the preserved points of the spectrum are selected by frequency. In other words, frequencies are determined as being useful to the diagnostics and they consequently serve as indicators through their coordinates in the complex plane, namely a real part, *Re(Z)*, and an imaginary part, *Im(Z)*. The electrochemical impedance spectra obtained contain about 40 frequencies. In this work, six frequencies are considered relevant, following expert observations [ONA 12, PAH 15a, SAN 15, WAS 10]. They are deemed as the most relevant and revealing of the shape of the impact of various failures.

Thus, as shown in Figure 4.10, 12 hyper-parameters should be extracted from the electrochemical impedance spectra, measured in various states of health of the fuel cell system.

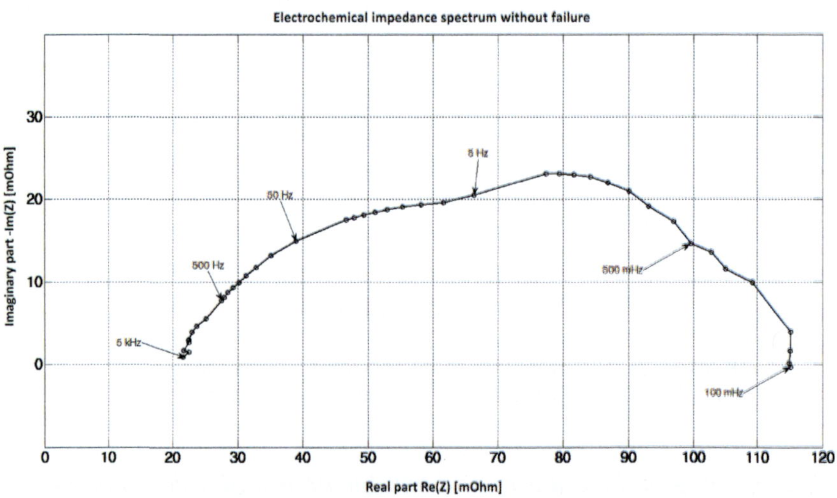

Figure 4.10. *Descriptors to be extracted from the EIS obtained using a stack of 40 cells*

Figure 4.11 shows the composition of databases of the FC with 40 cells. The various operating conditions are once again defined, associated with their electrochemical impedance spectra. The six characteristic frequencies are also provided and symbolized by f_x, where $x = \{1, 2, 3, 4, 5, 6\}$. For each of them, the pairs of coordinates {real part; imaginary part} are the training bases of the k-nearest neighbor algorithm.

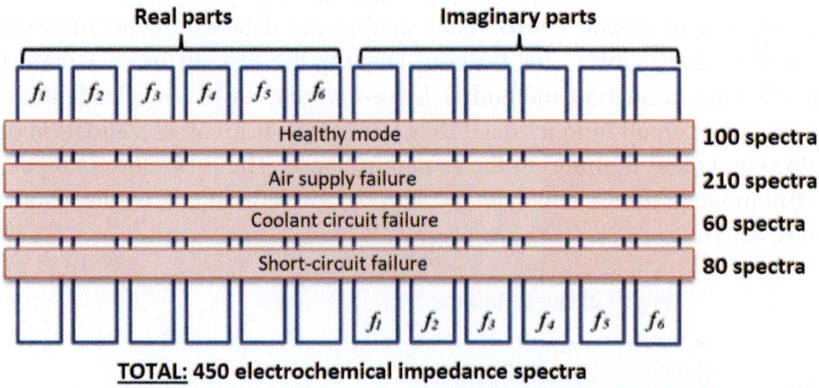

Figure 4.11. *Composition of the training base. For a color version of this figure, see www.iste.co.uk/jemei/cells.zip*

4.3.3.3. *Algorithm training and cross-validation method for the test*

The training phase is the algorithm's initialization stage, which brings the set of descriptors closer to a state of health or failure of the FC. The operating condition under which the EIS has been obtained is associated with it. In other words, each failure is labeled by 12 descriptors. This procedure is called data labeling. This amounts to associating a state of health or failure λ with a line of the learning matrix (Figure 4.12):

$$\left[\begin{matrix} Re(f_x)_{EIS_1} & Im(f_x)_{EIS_1} & Re(f_{x+1})_{EIS_1} & Im(f_{x+1})_{EIS_1} & \dots & Re(f_m)_{EIS_1} & Im(f_m)_{EIS_1} \\ & \vdots & & \vdots & & \vdots & \\ Re(f_x)_{EIS_n} & Im(f_x)_{EIS_n} & Re(f_{x+1})_{EIS_n} & Im(f_{x+1})_{EIS_n} & \dots & Re(f_m)_{EIS_n} & Im(f_m)_{EIS_n} \end{matrix}\right] \begin{matrix} \textit{state of health } \lambda_1 \\ \vdots \\ \text{state of health } \lambda_n \end{matrix}$$

Figure 4.12. *Learning matrix*

After the labeling phase, the algorithm learns the descriptor–state of health associations. It is only from this moment on that the classification of new individuals can be done. It is the test phase.

Given the extraction of 12 descriptors per spectrum, only a limited amount of points is available in case of failure. In order to avoid undermining the generalization ability of our classifier, cross-validation methods can be considered for evaluating the performances of the latter. A database can be divided into two parts: a sub-database is used for training, and once this is completed, a second database makes it possible to test the performances of the model. The best known among them include *k-fold*, *leave-one-out* or *Monte Carlo*. They enable the database to be subdivided into independent subsets for training and for the test. In these works, the retained cross-validation method is *leave-one-out*, as it is well adapted for optimal use of small number databases [BEN 04]. It involves validation on a single sample and training on the remaining base. The procedure is repeated for a number of times equal to the number of individuals in the database (Figure 4.13).

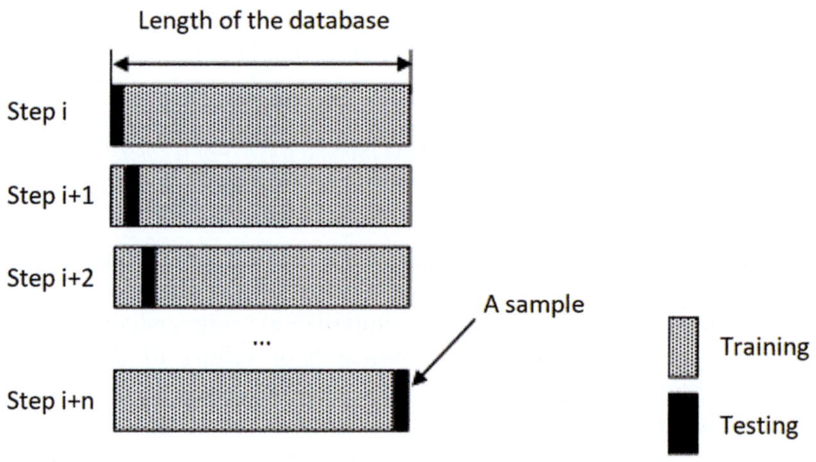

Figure 4.13. *Principle of "leave-one-out" cross validation*

4.3.3.4. *Classification results*

The proper classification results obtained on the stack of 40 cells with a value of k neighbors varying from 1 to 10 are summarized in Table 4.3. It is worth noting that the best classification rate is obtained with a number of neighbors $k = 2$. Increasing this parameter does not necessarily improve the results. This can be explained by the difficulty to separate various classes. Figure 4.14 is a perfect illustration of the weak interclass separability.

k	1	**2**	3	4	5	6	7	8	9	10
Rate (%)	84.44	**84.67**	83.56	84	83.78	84.22	83.56	84	83.56	83.33

Table 4.3. *Proper classification rate depending on k*

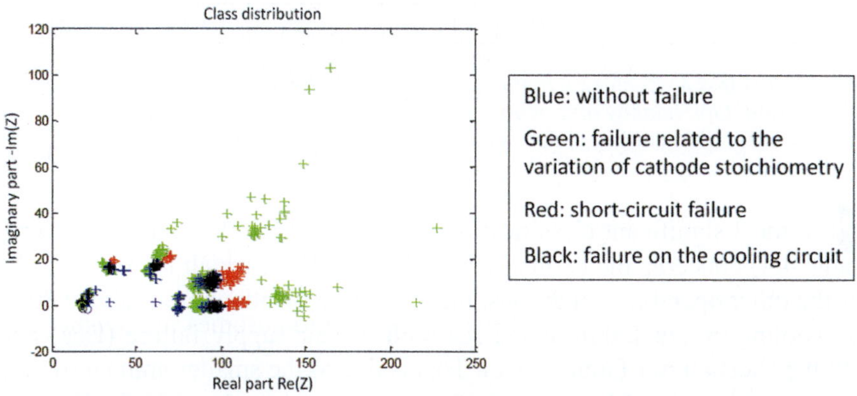

Figure 4.14. *Distribution in the complex plane of various electrochemical impedance spectra, obtained using the database of a stack of 40 cells after data labeling. For a color version of this figure, see www.iste.co.uk/jemei/cells.zip*

Table 4.3 shows that approximately 15% of the data are not correctly classified. In order to evaluate the labeling errors, a confusion matrix can be used. It enables us to compare the observed values of the dependent variable and those predicted, followed by a record of proper and improper predictions. Its representation in the form of a table enables us to display experimented states of health and the states of health diagnosed by a classification method. The ability of the algorithm to perform a proper classification with respect to the experimental data obtained can be evaluated at each line–column intersection. In theory, in case of perfect diagnostics, the diagonal of this table must be 100%. In practice, the more precise the diagnostics, the more the values of the diagonal tend towards 100%. The residues, namely the percentages corresponding to false estimations that are above the diagonal are also significant. Their interpretation enables us to correct the algorithm in order to improve the obtained performance and/or to determine what poses difficulties to the diagnostics.

		States of health estimated by the algorithm			
		$Opt_{estimated}$	$Dca_{estimated}$	$Dcr_{estimated}$	$Dcc_{estimated}$
States of health experimented	***Opt***	80%	10%	3%	7%
	Dca	0.5%	96%	1%	2.5%
	Dcr	1%	20%	55%	24%
	Dcc	1.5%	5%	10%	83.5%

Table 4.4. *Confusion matrix associated with the best classification rate. Opt: healthy and rated condition, Dca: air supply circuit failure, Dcr: cooling circuit failure and Dcc: short-circuit failure*

The most significant classification error obtained in the confusion matrix (Table 4.4) concerns the cooling circuit failure (Dcr), which is only of 55%. For the other operating conditions, the good classification rate is above 80%. The cooling circuit failure coincides with the air supply failure (Dca) and with the short-circuit failure (Dcc). This is due to the smaller amount of data related to this state of health (only 60 spectra compared to 200 for Dca, for example). A further reason can also be the difficulty to characterize this failure with the chosen game of descriptors.

In order to define the best game of descriptors, the choice has been made to conduct an exhaustive search of the best combination of descriptors to be extracted from the EIS. Exhaustive search is a tool that tests all the possible combinations of parameters. It relies on combinatorial logic and on a random selection without replacement (Figure 4.15). In our case, 4,095 possibilities are tested. Each pair of parameters is tested by the k-NN algorithm, and a good classification percentage is provided. It then remains to choose the best performance obtained and the associated combination of parameters. It is a solution composed of the following six descriptors: *Re*(500 Hz), *Re*(50 Hz), *Re*(5 Hz), *Re*(100 mHz), *Im*(500 mHz) and *Im*(100 mHz).

The good classification percentage obtained with this game of parameters reaches *92.22%* for k =1, which is significantly higher than the result obtained with 12 descriptors and $k = 2$ (84.67%). Moreover, if the associated confusion matrix is considered (Table 4.5), then it can be noted that residues are less significant.

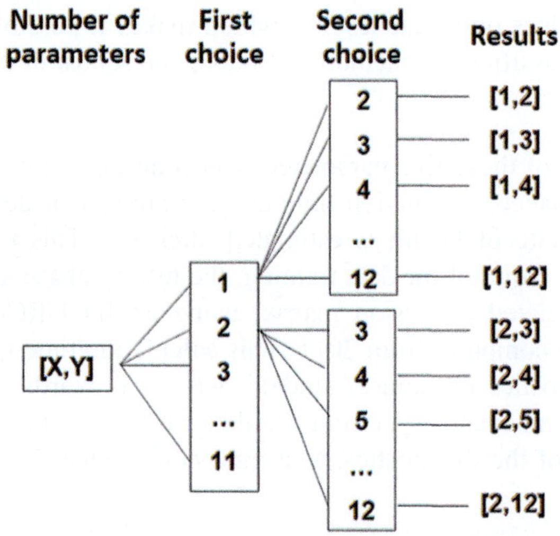

Figure 4.15. *Example of random selection without replacement*

		States of health estimated by the algorithm			
		$Opt_{estimated}$	$Dca_{estimated}$	$Dcr_{estimated}$	$Dcc_{estimated}$
States of health experimented	*Opt*	89%	10%	0%	1%
	Dca	7%	91.5%	1.5%	0%
	Dcr	0%	3%	93.5%	3.5%
	Dcc	0%	0%	1.5%	98.5%

Table 4.5. *Confusion matrix associated with the best classification rate with six descriptors*

This study has been realized on two other PEMFC of different sizes (8 and 20 cells) with other failures. The best classification rate (*93.66%* with k = 1) has been obtained for the stack of eight cells with only five descriptors. In order to generalize and transpose the developed method from one PEMFC to another, the five descriptors (*Re*(500 Hz), *Im*(500 Hz), *Im*(5 Hz), *Im*(500 mHz), *Im*(100 mHz)) resulting from the study on the stack of eight cells have been used on the stack of 40 cells. The good classification percentage reaches here *89.11%* for $k = 6$.

Performance loss only reaches 3%, which makes it possible to state that the supervised classification method is reliable and robust in terms of choice of descriptors.

The selection of these five parameters seems an excellent compromise in terms of performance, calculation time and the number of descriptors to be extracted. The state of health is estimated after 1 s. This time takes into account the time required for data training, the testing phase and the display of the best-associated confusion matrix, using an Intel RCoreTMi7-3770 CPU@3.40 GHz computer from 2011. This calculation time is sufficient for the online diagnostics of failures studied here. It is worth noting that the electrochemical impedance spectrum should be fully obtained for the proper implementation of the diagnostics, as a minimum frequency of 100 mHz is required.

4.3.3.5. *Transposition of the method for ageing*

The robustness of the method via descriptors extracted from the EIS has been shown in the previous sections. Here the aim is to transpose this method on another PEMFC in order to determine its ageing. Data have been obtained using a stack of five cells. The FC has operated for 2,000 h following a current mission profile, enabling the simulation of a micro-CHP system. During this long duration test, three EIS and a polarization curve are obtained per week.

The objective is to determine the increase in ageing during the 2,000 h of test following 10 cycles of micro-CHP per day (Figure 4.16). For the labeling phase, three states of health are chosen, namely (1) a state of beginning of life of the system, (2) a state of middle life and (3) a state of health of end of life of the fuel cell system. The end of the long-term test is correlated here with the system's end of life. Eight electrochemical impedance spectra are obtained in the first 1,200 operating hours, which are illustrated in Figure 4.17. It is possible to visually create groups, so that the first three spectra form the "beginning of life" group, the spectra t504 and t672 are dedicated to the "middle of life" and the last three are associated with the "end of life" label.

In order to test the algorithm, the five descriptors of the previous part will be chosen. The good classification rate reaches *87.5%* for $k = 3$, and the associated confusion matrix is given in Table 4.6.

The obtained results prove to be of high quality, as we are close to 90% of good classification.

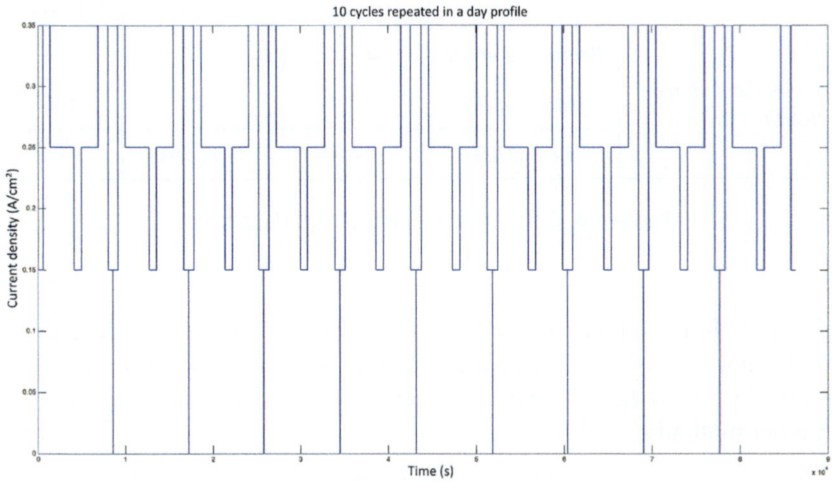

Figure 4.16. *Example of random selection without replacement*

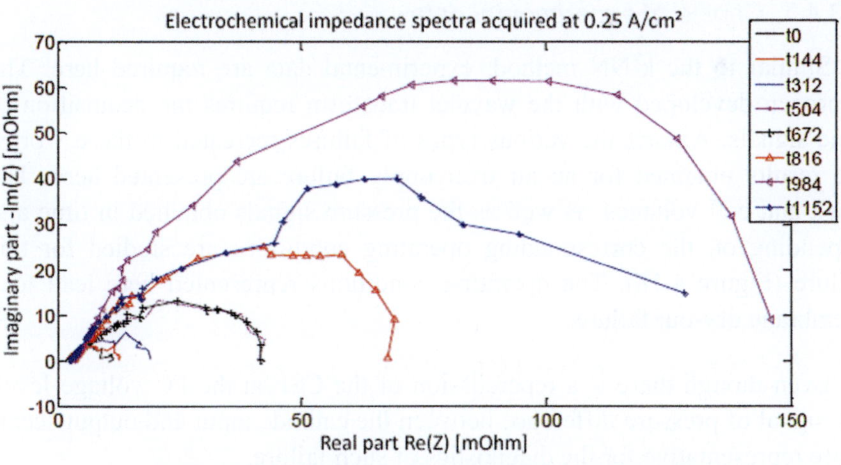

Figure 4.17. *EIS obtained during ageing. For a color version of this figure, see www.iste.co.uk/jemei/cells.zip*

		States of health estimated by the algorithm		
		***Beginning of life**$_{estimated}$*	***Middle of life**$_{estimated}$*	***End of life**$_{estimated}$*
Experimented states of health	***Beginning of life***	87.5%	12.5%	0%
	Middle of life	7%	100%	0%
	End of life	12.5%	12.5%	75%

Table 4.6. *Confusion matrix with five descriptors*

The k-NN method has been tested on various FCs and for the detection of various failures. Its high performances and robustness have been proven. The conclusion to this chapter provides an assessment and a comparison with other methods.

4.3.4. *Results obtained with the wavelet transform*

4.3.4.1. *Choice of experimental data*

Similar to the k-NN method, experimental data are required here. The approach developed with the wavelet transform requires the acquisition of time signals. Among the various types of failures recreated in these works, the results obtained for an air oversupply failure are presented here. The stack and cell voltages, as well as the pressure signals obtained in time and depending on the corresponding operating conditions are studied for this failure (Figure 4.18). The operating conditions represented here lead to a membrane dry-out failure.

Even though there is a repercussion of the CSF at the FC voltage level, the signal of pressure difference between the cathode input and output seems more representative for the diagnostics of such failure.

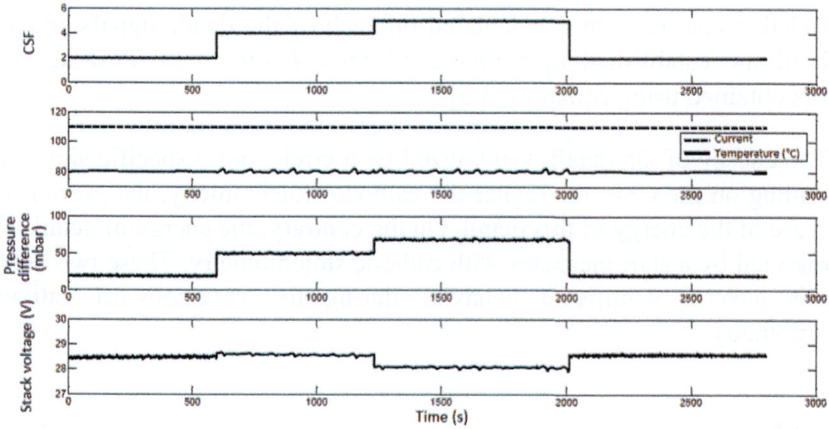

Figure 4.18. *Time evolutions of voltages, cathode (input–output) pressure difference and temperature of the FC depending on the CSF*

4.3.4.2. *Results of RWE*

The signal *s(t)* of the cathode (input/output) pressure difference is used for the validation of this method. The discrete wavelet transform is applied. Daubechies 4 wavelet is retained due to its efficiency, and a decomposition level 5 has been calculated using equation [3.11]. The results are presented in Figure 4.19.

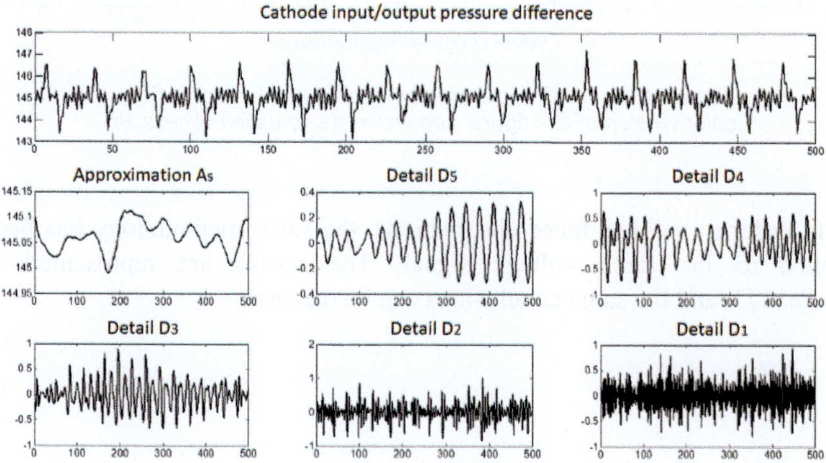

Figure 4.19. *Results of level 5 wavelet decomposition of the pressure signal*

Relative energy is then calculated on each of the detail signals and for each of the established operating conditions. Figure 4.20 illustrates the results obtained using equation [4.6].

The energy of *D5* detail, symbolized by a circle, has a specific signature depending on the CSF: the higher the cathode stoichiometry, the greater the decrease in the energy of this detail. On the contrary, the energy of detail *D1*, represented by a star, increases with cathode stoichiometry. These two detail signals have a significant behavior during the variations of cathode stoichiometry.

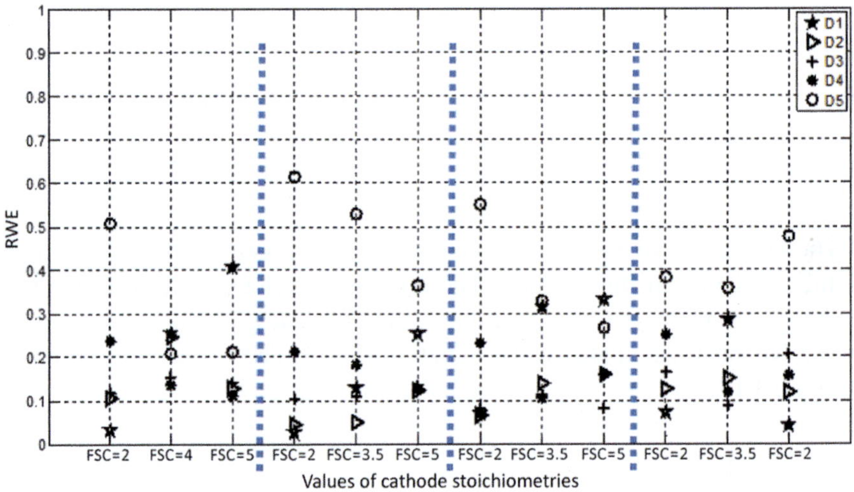

Figure 4.20. *RWE depending on operating conditions. For a color version of this figure, see www.iste.co.uk/jemei/cells.zip*

In order to validate these observations, the same methodology has been applied to the stack voltage signal. The results are represented in Figure 4.21, and the same conclusions can be reached.

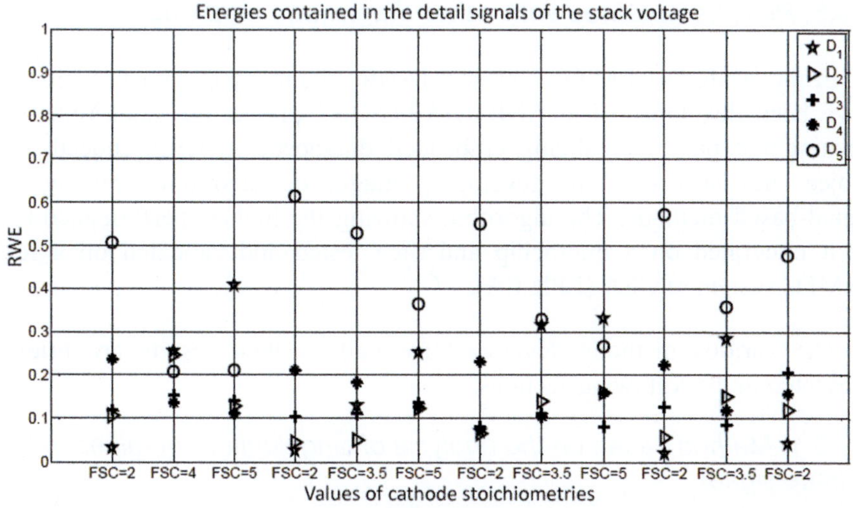

Figure 4.21. *RWE of the stack voltage depending on operating conditions*

The study of the energy indicator (RWE) on several detail signals obtained, by either voltage or pressure decomposition, reveals the signature of two extremum details. Indeed, regardless of the origin signal, the RWEs of details *D1* and *D5* react similarly to an increase in the cathode stoichiometry factor. The RWE of detail *D1* evolves as the CSF increases, while the RWE of detail *D5* shows the opposite behavior: when the CSF increases, its energy decreases. The other energies have neither a significant behavior nor a specific influence.

This RWE indicator can therefore be taken into account for establishing a diagnostics on the air oversupply failure in a PEMFC system.

The approach proposed here relies on works on SOFCs patented in 2009 [YOU 09b]. The method has been transposed to the PEMFC, which seems to offer encouraging results. It is worth noting that the same study involving the entropy of the same signals enabled the same conclusions to be drawn. Finally, to show the robustness of this method, it has been used for the detection and estimation of the ageing of a FC system. The results are once again encouraging [PAH 15b].

4.3.5. *Results obtained with other diagnostic methods*

In the ANR DIAPASON2 project [ANR 14], many diagnostic methods have been developed. The ANR DIAPASON2 project has contributed to opening the path to realizing embedded diagnostic systems. Indeed, the project has succeeded in developing diagnostic algorithms relying on signal-based methods. The algorithm showing the highest performance has been integrated on a microchip and then tested and validated on a real PEMFC system of CEA (EPICEA).

The various methods developed as well as their results are briefly presented in the following sections.

4.3.5.1. *Method based on the analysis of singularities contained in the voltage signal*

Multiscale analysis based on continuous wavelets has been used for observing the evolution of voltage signal across the complete stack over a wide range of scales. In order to quantify the force of singularities and their distribution for various operating conditions, the "multifractal formalism" based on the wavelet transform modulus maxima (WTMM) has been implemented. This method enables the discrimination of various operating conditions using singularity spectra calculated on the voltage with a single sensor [BEN 14a].

The results show that each failure imposes its own signature of singularities in the voltage. The best classification rate is obtained by combining the parameter selection technique MRMR and the k-NN classifier: *91.3%*.

4.3.5.2. *Statistical method*

This diagnostics relies on the study of cell voltages. A comparative study has focused on various possible couplings of various algorithms for parameter extraction and classification. For the extraction of parameters, four techniques have been considered: principal component analysis (PCA), Fisher discriminant analysis (FDA) and their nonlinear forms KPCA and KFDA. The following three algorithms have been used for the extraction: Gaussian mixture model (GMM), support vector machine (SVM) and k-nearest neighbors (kNN).

On the stacks, diagnostics has a confidence level above *95%*, and on the offline EPICEA system, it is above *90%*. Presenting the best compromise in terms of quality of results and ease of integration (calculation time less than 1 s), the combination of FDA and SVM has been retained for the validation [LI 16].

Finally, Figure 4.22 summarizes the various diagnostic methods developed within the thesis by Elodie Pahon and the ANR DIAPASON2 project.

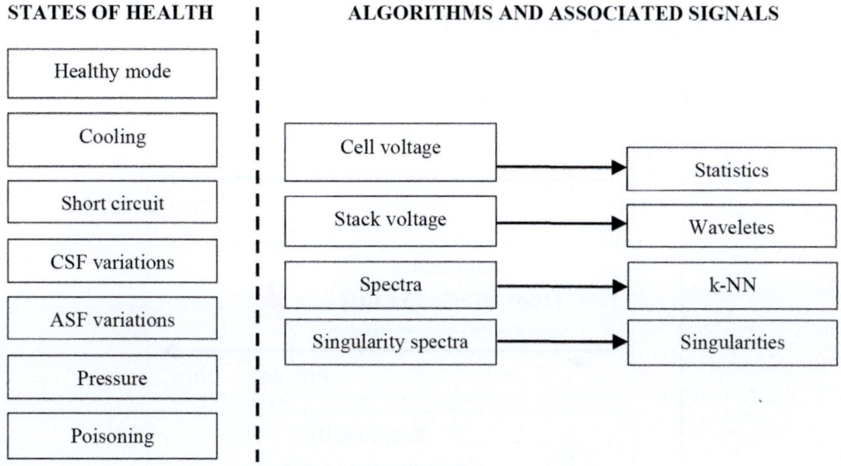

Figure 4.22. *Development of algorithms based on failures tested on PEMC benches*

Many works on diagnostics have been conducted for more than 10 years by our research team. Those presented here enable us to offer many perspectives to online FC diagnostics. This will contribute to increasing the reliability of current FC systems as well as their lifetime. Nevertheless, in order to drastically increase the lifetime of FCs, a new scientific and technological direction has emerged in recent years: prognostics of FCs. The following sections are dedicated to this subject.

4.4. Prognostics of fuel cells

The previous sections have highlighted the numerous methods used for the diagnostics of failures in a FC system. The methodology relies on the quantification of a failure indicator, obtained at a given instant on the system and its evaluation compared to those of rated operation. Significant deviations trigger the subsequent diagnostics (localization and identification) stage. On the contrary, prognostics enables the use of indicators reflecting the state of health of the system with sufficient time in advance, which enables us to make proper decision in time. In other words, diagnostics involves the analysis of the events preceding the malfunction (i.e. identification of the emergence of an event that had already occurred), while prognostics relates to the analysis of events before the occurrence of the malfunction (i.e. prediction of the future behavior of the system under observation) (Figure 4.23) [SAN 15].

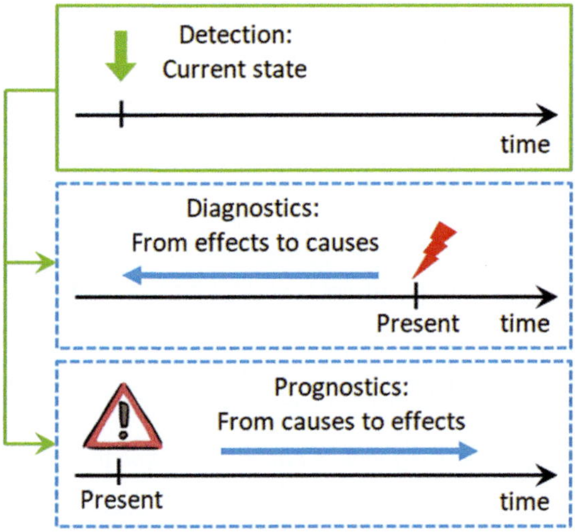

Figure 4.23. *Detection, diagnostics and prognostics. For a color version of this figure, see www.iste.co.uk/jemei/cells.zip*

Various projects and theses in our team aim to develop intelligent PHM (Prognostics & Health Management) methods for monitoring the state of health and estimating the remaining lifetime of PEMFC systems. The main scientific objective is the development of reliable prognostic approaches for the PEMFCs in various applications, and these aspects are described in the following sections. As a first step, the prognostics/PHM subject will be introduced and prognostic methods applicable to FC systems will be approached. Then in the second part, various prognostic methods developed by our team will be presented.

4.4.1. *From prognostics to PHM*

Traditional preventive and corrective maintenance methods are well known in the field of industrial systems. Nevertheless, they are often costly and constraining. In order to reduce the costs and the risks, manufacturers tend to increase their capacity to anticipate failures in order to resort to preventive actions as precisely as possible [GOU 15, HES 08, MUL 08]. For this purpose, "PHM – Prognostics and Health Management" is one of the increasingly used solutions and the prognostic process is today considered one of the main levers in the search for overall performance [GOU 15]. As proposed by the ISO committee, the aim of prognostics is "estimating the time to failure and the risk for one or more incipient failure modes" [ISO 04].

Prognostics is applied differently depending on the application and objectives considered. It is therefore important to have consistent evaluation criteria. For this purpose, prognostics can be evaluated in two ways:

– the main objective of prognostics is to provide the user with reliable data, which enables them to make proper decisions. Among these data, the most used metrics for prognostics is time to failure (TTF) or Remaining Useful Life (RUL). A confidence level should also be built in order to indicate the confidence value of RUL (Figure 4.24);

– it is also necessary to assess the quality of the prognostics in order to decide the adequate actions. The performance of the prognostic system should be established by quantifying, for example, the distance between a set of estimation of RUL and the exact (experimental) values of RUL.

These two metrics will be chosen in these works in order to qualify and quantify the FC prognostics. Before explaining the methods developed for realizing these prognostics, let us present the various available approaches to prognostics.

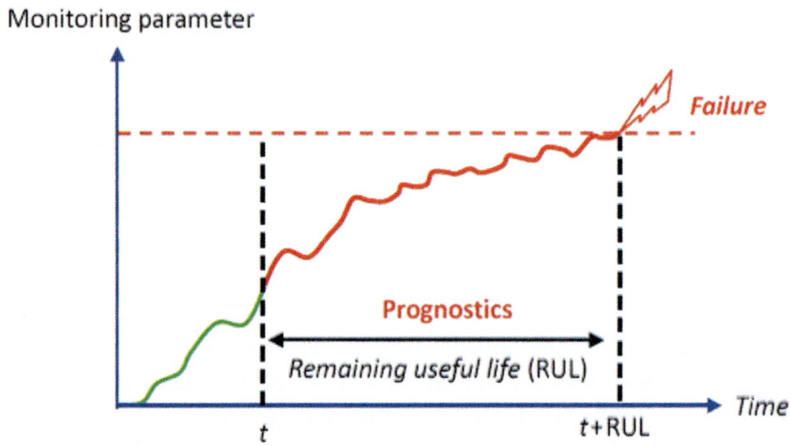

Figure 4.24. *Illustration of RUL. For a color version of this figure, see www.iste.co.uk/jemei/cells.zip*

4.4.1.1. *Various approaches to prognostics*

Four approaches are frequently used for setting up the prognostics of a system. They differ according to the application considered and to the available data or knowledge. They are summarized in Figure 4.25 and explained below:

– *Prognostics based on a physical model* uses mathematical representations in order to integrate a physical comprehension of the system's degradation process. This method provides precise results and does not require a history of data in order to estimate the future state once the model has been established. Nevertheless, the modeling of degradation is complex, as it does not involve the multiphysical and multiscale phenomena that need to be controlled. It is therefore difficult to obtain a dynamic model in the analytical form integrating the degradation phenomena. Moreover, once the model is built for a given application, it is very difficult to transpose it to another physical system. Their application framework is thus limited [JOU 15b].

– *Prognostics based on data* rely on the exploitation of monitoring data and use mainly the tools resulting from the field of artificial intelligence. This is a "black box" model and requires no in-depth knowledge of the system. The data obtained on the system are processed and enable the extraction of characteristics reflecting the behavior of the system and its degradation. A model is built thanks to the system's behavior training. It makes it possible to define the current State of Health (SoH) of the system and to predict its future. An estimation of the RUL can thus be formulated. The performance of this method is highly dependent on the quality and quantity of available data. Training these models can be critical in terms of both calculation time and the real representation of the system's behavior. Nevertheless, this approach offers an excellent compromise between its applicability and the precision of the results obtained [GOU 15, JOU 15a, MOR 15a, SAN 15].

– *Prognostics based on experience* rely on the exploitation of knowledge obtained through the experts or through feedback over significant time. This knowledge enables the use of statistical tools and reliability functions that translate the system's failure evolution. This approach requires neither analytical models representing the physical behavior of the system nor in-depth knowledge of the degradation mechanisms. The main drawback is that it is difficult to have a history of past experience that is representative for all the conditions of use of the system. The results obtained are less precise than those based on a model or on the data, particularly when the operating conditions are variable or when the studied technology is recent [SOU 13].

– *Hybrid prognostics* is a method that integrates a physical behavior model and a data-oriented approach. Experimental data are used here both for the estimation of non-observable parameters of the analytical model and for the estimation of phenomena that are difficult to model. Hybrid approaches show good estimation and prediction performances. Moreover, they enable good quality modeling of uncertainties. However, they can be very demanding in terms of calculation resources, and are limited by the need for a physical model of degradation phenomena [JOU 15b].

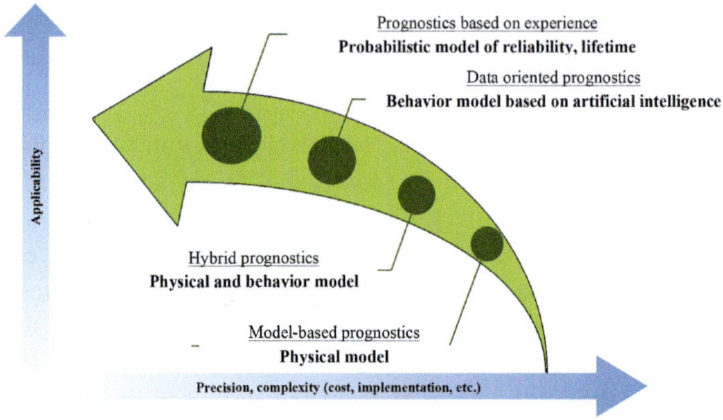

Figure 4.25. *Classification of prognostic methods. For a color version of this figure, see www.iste.co.uk/jemei/cells.zip*

4.4.1.2. *PHM*

Regardless of the retained prognostic approach, it is worth noting that this process cannot be dealt with in an isolated manner. Indeed, prognostic implementation requires a set of tasks (or modules) that are commonly regrouped under the term PHM. PHM is the extended view of condition-based maintenance (CBM) [ISO 06] whose architecture is represented in Figure 4.26.

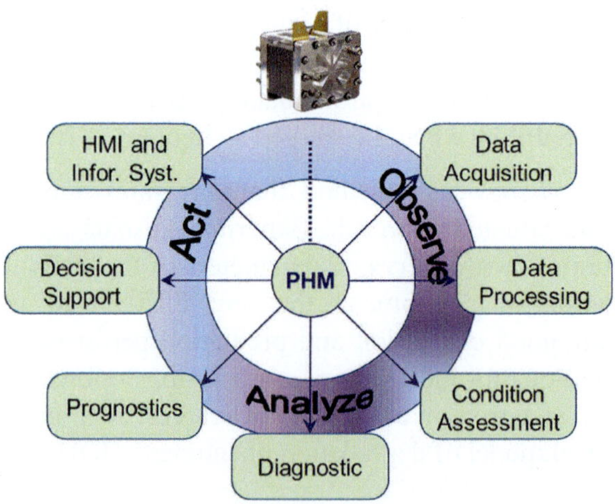

Figure 4.26. *PHM modules. For a color version of this figure, see www.iste.co.uk/jemei/cells.zip*

The various stages are presented as follows:

– *data acquisition* aims to collect numerical data issued by the sensors;

– *preprocessing* aims to filter the collected signals and extract/select the descriptors, enabling the characterization of the operation;

– *monitoring* aims to compare the descriptors with expected values in order to generate alerts depending on predefined thresholds;

– *diagnostics* aims to determine if the state of the system is degraded or not, and to suggest probable failure mechanisms and causes;

– *prognostics* aims to predict the future state of the system and estimate the time to failure;

– *decision aid* aims to provide recommendations regulating actions for the accomplishment of system's mission;

– *human–machine interface* aims to present the results.

These works have, until now, focused on various points enabling the correct understanding of PEMFC degradations (data acquisition, extraction of descriptors, detection, diagnostics). The present task is to define a healthy operation lifetime of a FC and to eventually apply the decisions that enable the best operation of this FC.

The subjects of prognostics and PHM are under rapid expansion within the framework of predictive maintenance. They are perfectly well adapted for improving the FC systems. The use of these tools in the development of FCs is nevertheless quite recent. To our knowledge, the first works on the prognostics of the FC, at the international level, have started in 2011 as part of a German project financed by the German Ministry of Economy and Technology. At the national level, the ANR PROPICE project has been the first project dedicated to the prognostics of FC. In 2012, only one article [ZHA 12] predicting the loss of active surface could be found in the literature. In this article, prognostics was realized from an extension of the Kalman filter. The literature has enriched since 2013, thanks to the works launched by the research federation FCLAB on the FC prognostic tools. Indeed, hybrid approaches [JOU 13, JOU 14a, JOU 14b, JOU 15a, JOU 15b, JOU 16], based on models [LEC 15] or data-oriented approaches [MOR 14,

MOR 15a, MOR 15b, SAN 14a, SAN 15] have emerged. Other data-oriented approaches have also been proposed within IEEE PHM Data Challenge [PHM 14] organized by FR FCLAB in order to predict the state of health based on impedance measurements [KIM 14, VIA 14] as well as to estimate the RUL based on voltage data [HOC 14, KUR 14].

As can be noted throughout this book, one of the areas of specialization of our research team is the experimental testing of PEMFC. Thanks to numerous data collected in recent years and to the experience gained on methods based on artificial intelligence and particularly on neural networks, it was obvious that the prognostic approaches developed would be data oriented. This will be the focus of the following sections.

4.4.2. *Prognostic methods developed*

Only the data-oriented methods developed in the theses of Rosa-Elvira Sanchez [SAN 15] and Simon Morando [MOR 15a] are presented here. The method developed in the thesis by Rosa-Elvira Sanchez relies on an ANFIS (Adaptive Neuro-Fuzzy Inference System) approach, while the works of Simon Morando are based on neural networks specific architecture.

4.4.2.1. *ANFIS-based method [SAN 15]*

These works propose a model to predict the state of health of the FC. The methodology relies on the time evolution of the voltage of the FC. This simple and inexpensive indicator is used to monitor the current state of health (SoH) of the stack, which is a good indicator of the degradation of the FC. Thus, the prediction of the time evolution of voltage can serve to provide a good estimation of future degradation or failure.

It has been shown in Chapter 3, and particularly in Mona Ibrahim's thesis [IBR 13], that the prediction of a time series could be obtained by means of various strategies and models. A recursive strategy will be retained here, as it can be easily implemented and offers a good compromise between precision and complexity. Furthermore, prediction requires the establishment of only one model, which enables a reduction of calculation times [SAN 15]. In terms of model, the ANFIS (Adaptive Neuro-Fuzzy Inference System, Figure 4.27) systems are of interest to us, as they are rapid, adaptive and do not require a complex analytical model. Moreover, they can be easily implemented in real-time [GAO 01, NAU 99, WU 08c]. It is worth noting in the context of our study that neuro fuzzy systems enable the modeling of nonlinear systems, such as the FC, and are able to take into account human experience. Moreover, it is possible to integrate measurement noises of training data.

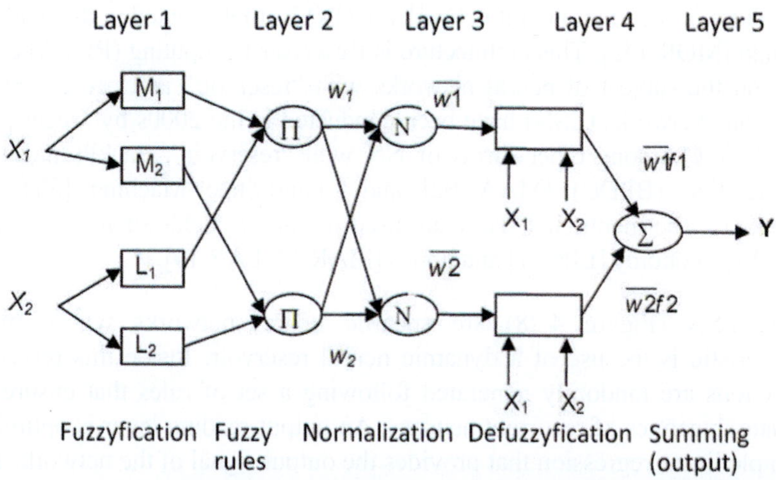

Figure 4.27. *Architecture of an ANFIS model*

It should be recalled that, as already explained in Chapter 3, fuzzy logic is a calculation technique that uses linguistic labels to represent numerical values. The systems based on fuzzy logic rely on a collection of "if–then" and membership functions (MF) that define the relationship between the system's input and output variables. Fuzzy rules and MFs are defined using either human knowledge or the system's data that define the relationship between input and output variables. Given the complexity and the lack of expertise in the estimation of the lifetime of a PEMFC, the design of a fuzzy system for predicting the loss of performance depending on human experience is difficult. Consequently, the design of such a system must involve the use of experimental data.

ANFIS systems are a class of adaptive networks introduced by Jang et al. [JAN 93], which can be considered as a structure of neural network with direct action, in which each layer is a component of the neuro-fuzzy system. In other words, these systems rely essentially on fuzzy logic in which the rules and the MFs are optimized by the intermediary of a neural network. The use of this architecture will be explained in section 4.4.3.2.

4.4.2.2. *ESN-based method [MOR 15a]*

Neural networks (NN) are particularly well adapted to nonlinear, dynamic systems, when the physical parameters of the concerned system are difficult to measure. These tools are interesting for the study of FCs [JEM 04]. A specific architecture of recurrent neural networks (RNN) is implemented in the works of Morando [MOR 15a]. This architecture is Reservoir Computing (RC). The first works on the subject of neural networks with "reservoir" and particularly on Echo State Networks (ESN) have been conducted in the 2000s by Jaeger [JAE 01, JAE 02]. Among other forms of NN with "reservoir", BackPropagation-DeCorrelation (BPDC) [MAA 02] and Liquid-State Machine [STE 04] networks can be mentioned. They are used in various fields such as medicine [ONG 13], economy [LIN 11] and optics [MAR 12, LAR 10].

The ESN (Figure 4.28) are specific neural networks whose major characteristic is the use of a dynamic neural reservoir. Inside this reservoir, connections are randomly generated following a set of rules that ensure the adequate dynamics of recurrent neurons. An output-reading layer is optimized by simple linear regression that provides the output signal of the network. This new approach benefits from inherent capacities of time processing of recurrent networks, while simplifying the method for network formation.

Moreover, the training of such a network is simplified, as the weights at reservoir output are only optimized. This necessarily renders the algorithm simpler and much more rapid.

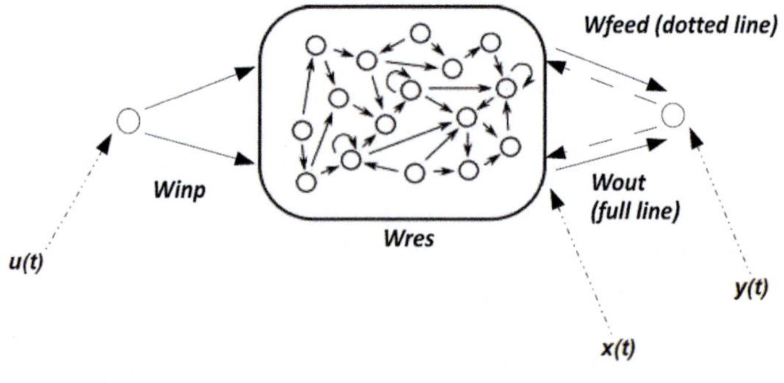

Figure 4.28. *Architecture of the echo state network. N_{res} is the number of neurons in the reservoir, K is the number of inputs, L is the number of outputs, W_{inp} is the matrix connecting the inputs to the reservoir (N lines, K columns), W_{res} is the matrix representing the reservoir (N lines, N columns), W_{out} is the matrix connecting the reservoir to the output of the 'ESN (L lines and (N+K) columns) and W_{feed} is the matrix representing the feedback of the output on itself (N lines, L columns)*

Even though the results obtained with the ESNs are impressive in the field of time signal processing [JAE 03, VER 07], the parameterization of this type of tool is far more complex than a classical neural network.

Optimization of the parameters of the reservoir (listed below) is required:

– The number of neurons, N_{res}, in the reservoir is an essential parameter, as it will partly define the performance of the network. Theoretically, obtaining a better linear combination corresponding to the best possible solution requires defining a reservoir with a substantial number of neurons. However, there is a risk of facing overfitting and the network will no longer have the capacity to generalize. Moreover, the use of a too large number of neurons causes an increase in the calculation time. It is therefore necessary to reach a compromise between precision and calculation time.

– Connectivity, c, represents the percentage of weights of non-null values in the reservoir. It ranges between 0 and 1. Indeed, in order to obtain better results, the neurons of the reservoir should not be all interconnected.

– Spectral radiation, $\rho(W_{res})$, enables a scaling of the matrix W_{res}. More specifically, it is the scaling of the width of the distribution of non-null elements of W_{res}.

– The leakage rate of the reservoir, α, can be considered as the speed of the dynamic update of the reservoir.

In addition to these parameters, the retained training algorithm and the topology of the reservoir should be defined. These subjects are presented in the following sections.

4.4.3. *Results obtained with ANFIS*

The prediction model is developed using ANFIS, and the time variation of voltage is one of its inputs. Indeed, voltage reflects the degradation of the FC performance. This degradation has been studied through long-term tests covering 1,000 h of continuous operation (24/7) each. In the first part, the data used for establishing the model are presented, then in the second part, its architecture is defined and in the end, the results obtained are presented.

4.4.3.1. *Data acquisition and preprocessing*

Long-term tests have been conducted on two PEMFCs of five cells with an active surface of 220 cm². The two FCs, denoted FC1 and FC2, were subjected to two different tests. A constant current of 110 A has been applied to FC1 for 1,000 h with an operating temperature of 75°C, ASF = 1.5 and CSF = 2 (these parameters are considered constant throughout the test). For FC2, the same parameters have been retained for the test, with the only difference being that a dynamic triangular stress of frequency 5 kHz and amplitude $\Delta I = \pm 10\%$ has been superimposed to the continuous component of 110 A. This evidences the behavior of the FC when it is connected to the DC/DC converter. The results obtained for FC1 and FC2 are shown in Figures 4.29 and 4.30.

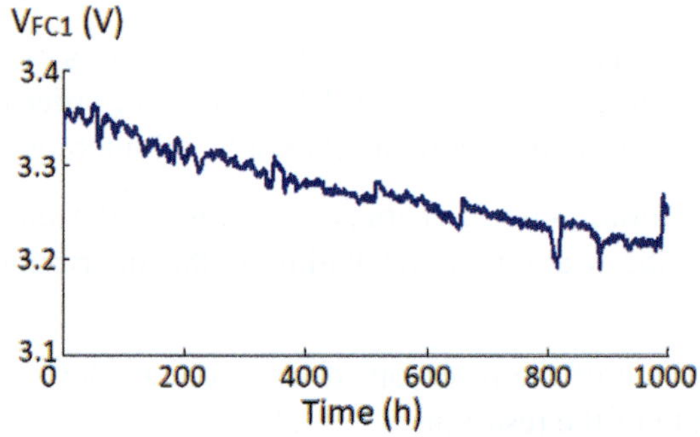

Figure 4.29. *Time evolution of voltage across FC1*

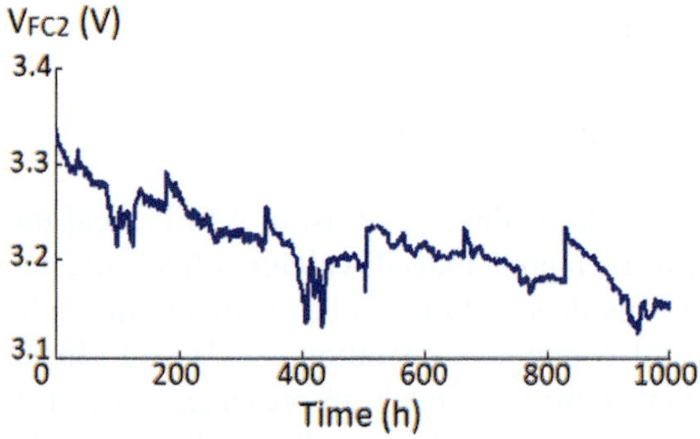

Figure 4.30. *Time evolution of voltage across FC2*

Time evolutions of voltages have been obtained with a frequency $f_s = 1$ Hz, which represents a database composed of 3,600,000 measurement points per stack. To avoid the training time of the algorithm being weighed down by too heavy databases, a study has been conducted to reduce this base while preserving the representative information on degradation. This has led to the conclusion that a sampling of one point per hour is a good compromise between the calculation time and the quality of training.

One objective during these tests is the study of the natural ageing of the FC. Various disturbances can lead to performance drops or even to performance regains that are completely decorrelated from the natural ageing of the stack. These disturbances can be of various natures. There may simply be periodic characterizations (polarization curves or EIS), enabling the estimation of the state of health of the FC, inadvertent stops of the test bench or of the installation or even a deviation from the operating point of the FC. In such cases, it would be difficult for the prediction model to take into account these external disturbances.

Our proposal is therefore to filter the time signal of the voltage in order to obtain a more or less smooth signal of the stack voltage and a component reflecting only the external disturbances. This also makes it possible to have a tendency that facilitates data projection over a future horizon. Figure 4.31 illustrates the results obtained. It is obvious that filtered voltage will be used as the reference voltage for the model, particularly during the training phase.

With the database being established, the architecture of the model should now be examined.

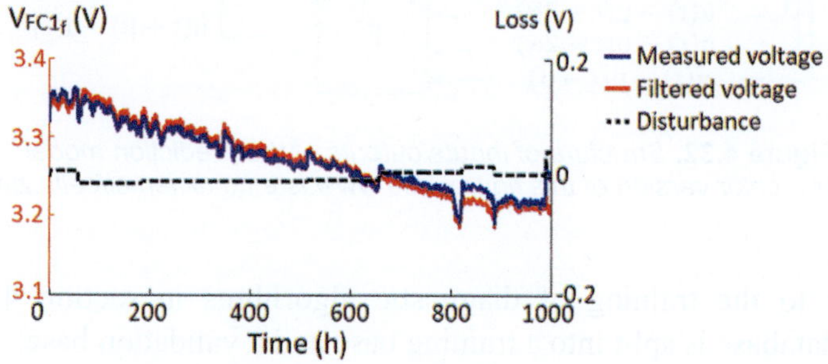

Figure 4.31. *Decomposition of the reference signal into two components for stack no. 1. For a color version of this figure, see www.iste.co.uk/jemei/cells.zip*

4.4.3.2. *Definition of the model*

Establishing the prediction model involves several stages. First, it is essential to choose the proper inputs/outputs and then to realize the training in order to obtain a prediction. Finally, the performance and the relevance of results should be evaluated.

The input of the model considers a set of N values including the current value and $N-1$ regressors of the time series represented by $u(t)$. Voltage drop is taken into account (equation [4.7]). Figure 4.32 represents the retained structure with five inputs, one output and three regressors:

$$\hat{u}(t+\beta)-u(t)=F\left[u(t)-u(t-(N-1)\alpha),\ldots,\, u(t)-u(t-\alpha)\right] \quad [4.7]$$

where α is the time between each regressor and β is the time between the current value and the predicted value.

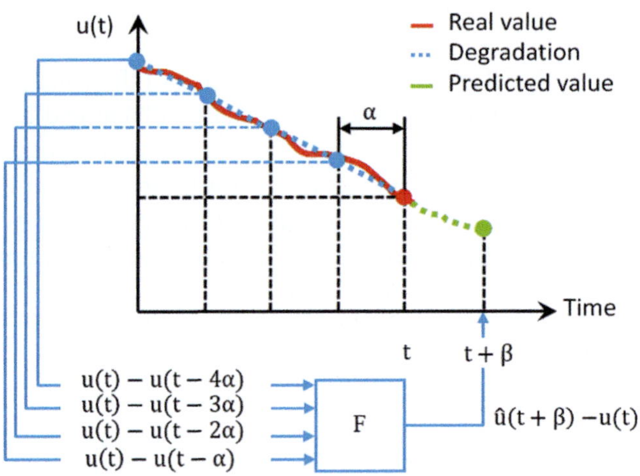

Figure 4.32. *Structure of inputs/outputs of the prediction model. For a color version of this figure, see www.iste.co.uk/jemei/cells.zip*

Similar to the training of diagnostic algorithms in section 4.3.3, the available database is split into a training base and a validation base.

The first set of observations therefore enables the training and the second set is used to validate the prediction. N-step prediction is realized by an iterative approach. This approach uses a unique model that is tuned to make an estimation of "one-step ahead prediction" $\hat{u}_{t+1}$. The estimated values are used as independent variables of the model in order to estimate the future outputs in time. The operation is repeated until the estimation of the signal value reaches the prediction horizon $\hat{u}_{t+H}$. The main drawback of this approach is error propagation: precision decreases as the prediction horizon length increases.

An iterative structure coupled with the ANFIS model appears to be an interesting method. Nevertheless, one of the difficulties is to define the game set of parameters, enabling the best compromise between the calculation time and precision.

To provide the response elements, several sets of parameters have been considered for the prediction structures involved (Table 4.7). The best results are presented in the following section.

Control parameters	***Values***
Number of inputs	$N \in [3,5]$
Number of outputs	1
Type of MF	Gaussian
Number of MFs	$MF \in [2,10]$
Training data	$trn_{data} \in [100, 800]$
Time gap between values	$\alpha \in [1, 20]$

Table 4.7. *Parameters of the model used*

4.4.3.3. *Simulation results*

The results obtained are studied by quantifying the prediction performances. For this purpose, standard statistical criteria, such as the root mean square error (RMSE, equation [4.8]), mean absolute percent error (MAPE, equation [4.9]) and the coefficient of determination (R^2, equation [4.10]), are used:

$$\text{RMSE} = \sqrt{\frac{1}{m}\sum_{i=1}^{m}(\hat{y}_i - y_i)^2} \qquad [4.8]$$

$$\text{MAPE} = \frac{1}{m}\sum_{i=1}^{m}\frac{|\hat{y}_i - y_i|}{|y_i|}\times 100 \qquad [4.9]$$

$$R^2 = 1 - \frac{\sum_{i=1}^{m}(y_i - \hat{y}_i)^2}{\sum_{i=1}^{m}(y_i - \overline{y_i})^2} \qquad [4.10]$$

where $\hat{y}_i$ is the expected value, y_i is the real value, $\overline{y_i}$ is the mean value of the observed set and m is the number of observed values.

The RMSE and MAPE values must be close to zero, while a coefficient of determination close to the unit indicates a good adjustment of the prediction model.

Figures 4.33–4.36 show the best predictions obtained with the set of parameters $N = 4$, $MF = 3$, $\alpha = 3$ for V_{FC1}; $N = 4$, $MF = 3$, $\alpha = 4$ for V_{FC1f}; $N = 4$, $MF = 2$, $\alpha = 11$ for V_{FC2}; $N = 4$, $MF = 2$, $\alpha = 3$ for V_{FC2f}. During the training phase, the amount of data used is $t_0 = 0$ h to $t_1 = 500$ h over a prediction horizon $H = 500$ h.

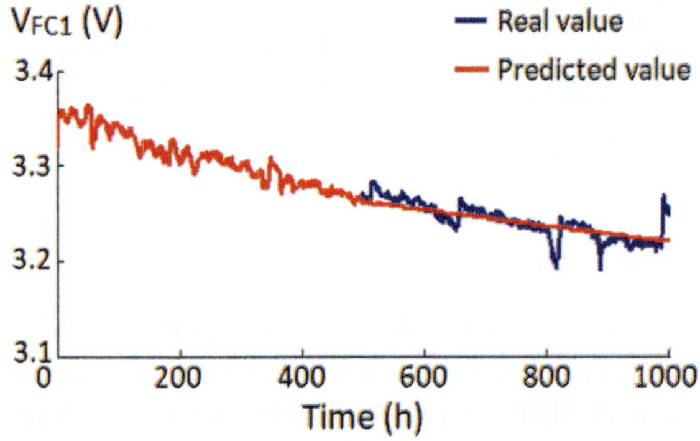

Figure 4.33. *Prediction results for FC1. For a color version of this figure, see www.iste.co.uk/jemei/cells.zip*

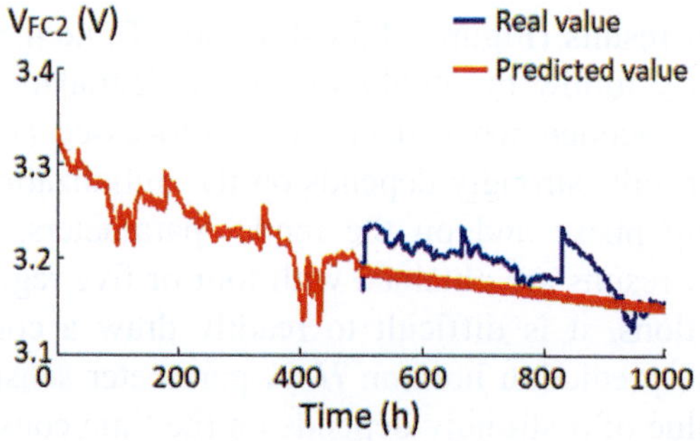

Figure 4.34. *Prediction results for FC2. For a color version of this figure, see www.iste.co.uk/jemei/cells.zip*

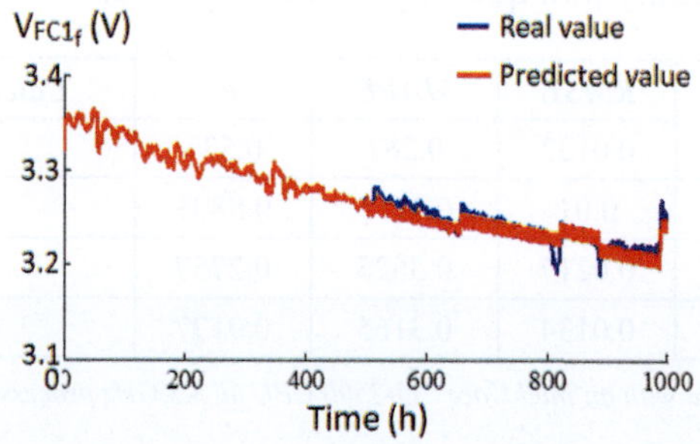

Figure 4.35. *Prediction results for FC1 with filtered input voltage. For a color version of this figure, see www.iste.co.uk/jemei/cells.zip*

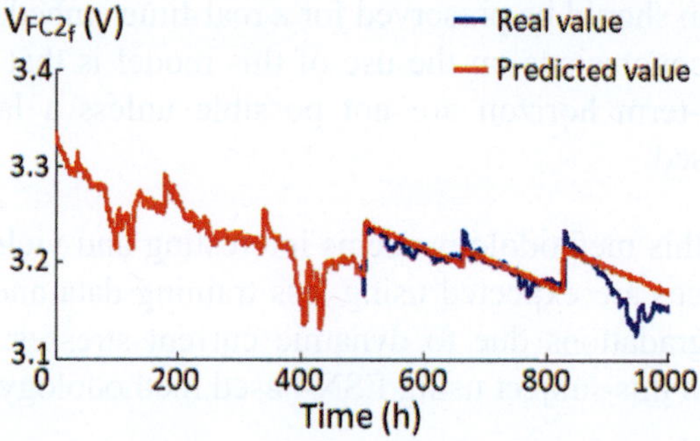

Figure 4.36. *Prediction results for FC2 with filtered input voltage. For a color version of this figure, see www.iste.co.uk/jemei/cells.zip*

The simulation results (Figures 4.33–4.36 and Table 4.8) show that the model can perfectly follow the evolution of the degradation of a PEMFC. Better results are obtained when data are preprocessed (Figures 4.35 and 4.36). The set of results strongly depends on the initialization of parameters during the training phase and on the model parameters. While it seems obvious that better results are obtained with four or five regressors and three membership functions, it is difficult to readily draw a conclusion for the value of α and the prediction horizon H. A parameter sensitivity study has shown that the value of α strongly depends on the time constants and on the degradations involved. Indeed, good sampling is required in order to avoid losing information in the signal or conversely having too much data and significantly increasing the calculation times. As for the prediction horizon, it depends on the quality and quantity of training data.

Predicted signal	*RMSE*	*MAPE*	R^2	*Calculation time (s)*[1]
V_{FC1}	0.0123	0.281	0.535	1
V_{FC1f}	0.01	0.2455	0.8891	3.15
V_{FC2}	0.0217	0.5625	0.2757	0.144
V_{FC2f}	0.0134	0.3165	0.9177	2.146

1 Calculation time obtained with an Intel Core™i5-2500 CPU @ 3.3 GHz processor

Table 4.8. *Prediction results*

It is important to keep in mind that a compromise between calculation time and precision should be preserved for a real-time embedded application. One of the first conclusions on the use of this model is that the predictions aimed at a long-term horizon are not possible unless a large amount of training data is used.

Even though this methodology seems interesting and yields good results, better performances are expected using less training data and being capable of predicting degradations due to dynamic current stresses. The following section focuses on this subject using ESN-based methodology.

4.4.4. *Results obtained with ESNs [MOR 15a]*

The works conducted in the thesis by Simon Morando rely on data-based prognostic methods. The predictive tool used here is an architecture specific to Neural Networks, namely Reservoir Computing. The results obtained so far are very encouraging, enabling a prediction of cell voltages until the end of lifetime of FC with an error below 5%.

As the first step, the data required for network training are presented, followed by a detailed presentation of how the architecture has been designed, and also of the optimization of its parameters, and then by a presentation of the results obtained with this methodology.

4.4.4.1. *Choice of databases*

The databases used in the set of results obtained with ESNs stem from two projects. The regional project Burgundy Franche-Comté (UBFC) PHM-FC has enabled us to conduct a long-term experimental campaign of a PEMFC with a constant load profile, while the European project SAPPHIRE has provided ageing data issued from a dynamic profile. Time evolutions of voltages are essentially presented here, as well as several characterizations which make it possible to define the state of health of the FC.

The tests conducted in the UBFC PHM-FC regional project correspond to the time evolution of the voltage of the FC responding to a constant load profile (current density of 0.6 A/cm^2), with a series of quasi-static and dynamic characterizations (polarization curves U(I) + EIS) obtained every 168 h. In particular, this test enables the observation of the natural ageing of this PEMFC, with five cells having an active surface of 100 cm^2. The results are presented in Figure 4.37. It is worth noting that an installation cutoff of approximately 1,700 operating hours has forced us to modify the operating parameters in order to maintain an acceptable overall voltage.

Many voltage jumps can be noted in Figure 4.37. They correspond to weekly characterizations (Figures 4.38 and 4.39). To avoid biasing the network training, they are manually removed (Figure 4.40).

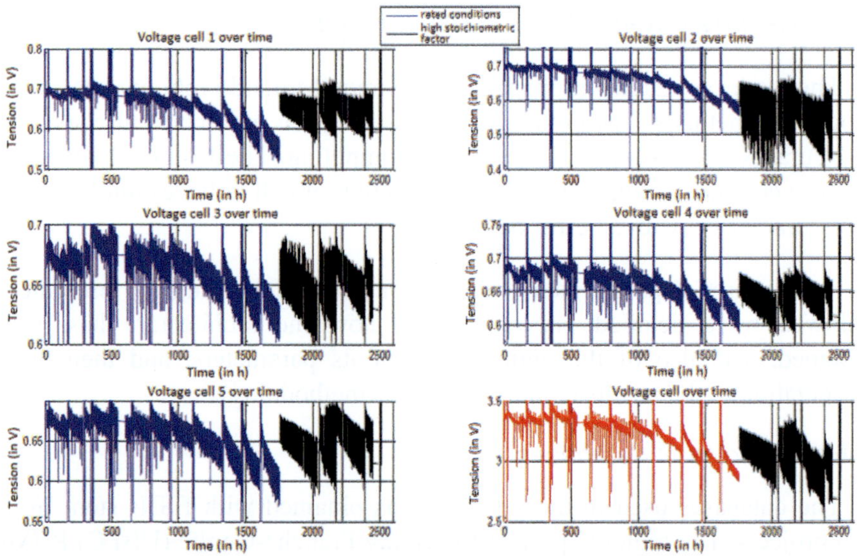

Figure 4.37. *Evolution of cell and stack voltages in the regional project. For a color version of this figure, see www.iste.co.uk/jemei/cells.zip*

Subsequently, in order to have a more substantial database, the five individual cell voltages are considered as five single cell fuel cells subjected to the same profile, with characterizations in the same days. For the training, four cells are averaged for predicting the evolution of the fifth.

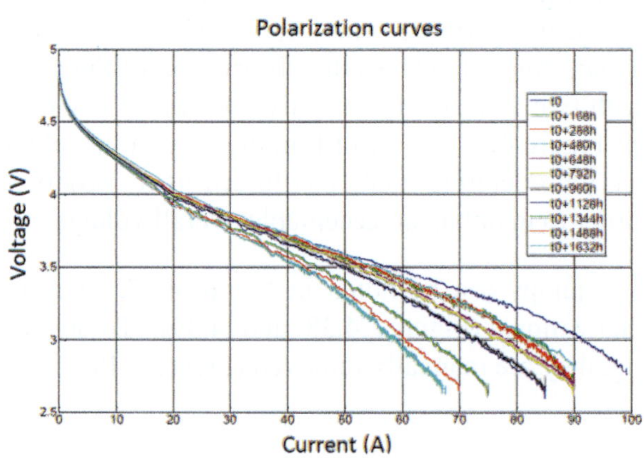

Figure 4.38. *Polarization curves at various instants of life. For a color version of this figure, see www.iste.co.uk/jemei/cells.zip*

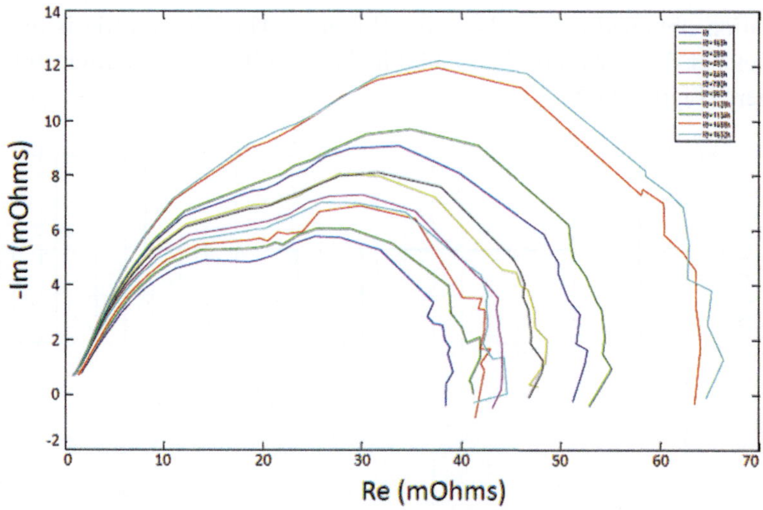

Figure 4.39. *EIS at various instants of life. For a color version of this figure, see www.iste.co.uk/jemei/cells.zip*

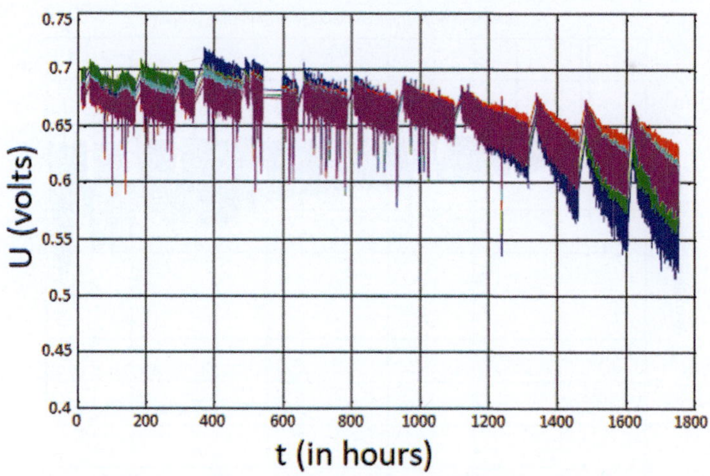

Figure 4.40. *Evolution of filtered cell voltages. For a color version of this figure, see www.iste.co.uk/jemei/cells.zip*

One of the objectives of the experimental campaign of the European project SAPPHIRE is to find the time evolution of the voltage of the PEMFC when it follows a profile of dynamic current. It is the profile of a stationary application obtained over a 24 h day (Figure 4.41). This profile is repeated 10 times per day for more than 2,000 h. The results obtained after 600 h are used here (Figure 4.42).

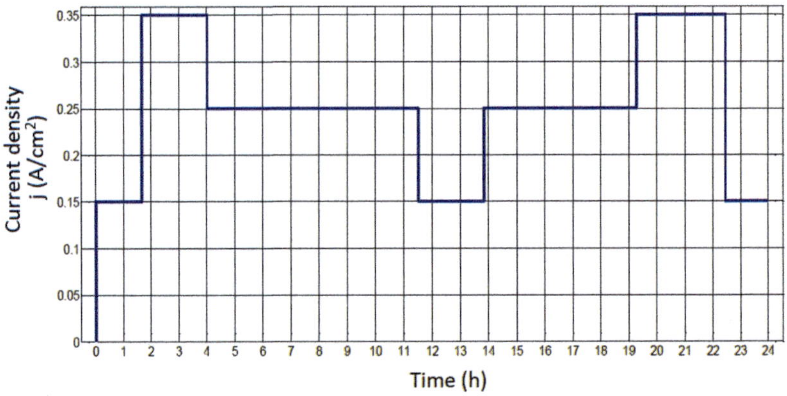

Figure 4.41. *Current profile in SAPPHIRE project*

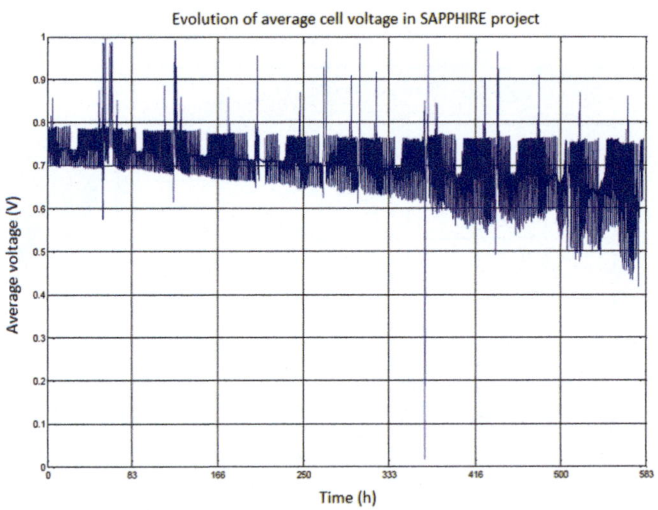

Figure 4.42. *Time evolution of voltage in SAPPHIRE project*

Similar to the regional project, a characterization series has been realized every 168 h. The results are shown in Figures 4.43 and 4.44.

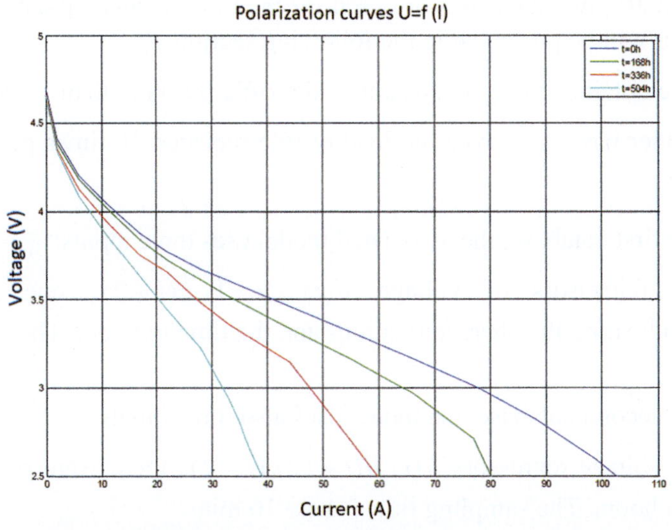

Figure 4.43. *Polarization curves at various instants of life. For a color version of this figure, see www.iste.co.uk/jemei/cells.zip*

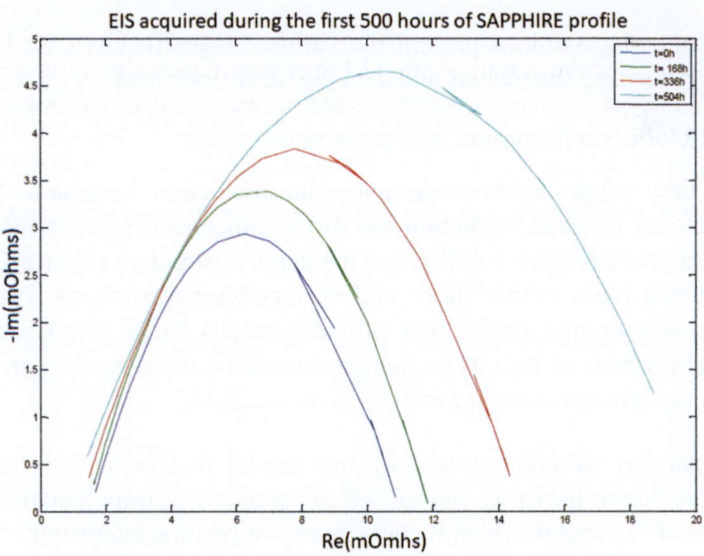

Figure 4.44. *EIS at various instants of life. For a color version of this figure, see www.iste.co.uk/jemei/cells.zip*

In summary, two databases are used for the training and validation of the models that will be presented in the following section:

– a base obtained using a constant load profile for more than 1,700 h;

– a further base on a dynamic load profile repeated 10 times per day for 600 h.

For the first database, the associated model uses three inputs:

– three regressors of voltage $u(t), u(t-1), u(t-2)$, one Boolean operator indicating the characterizations and the time in hours. The sampling time is one hour.

For the second database, the model uses also three inputs:

– three voltage regressors $u(t), u(t-1), u(t-2)$, the current profile and the time in hours. The sampling time is here 10 min.

Finally, since various data have various scales (from 0 to 1.2 V for the voltages and from 0 to 1,700 h for the time), they will be normalized between 0 and 1 so that they have all the same influence on training.

The following sections present the various stages required for building and parameterizing the models and then the results obtained.

4.4.4.2. *Model construction and parameterization*

The first stage involves defining the prediction structure. Various structures can be used, including the direct structure (Figure 4.45(a)), the parallel structure (Figure 4.45(b)) and the iterative structure (Figure 4.45(c)). The first two types enable short- and average-term predictions. The direct structure is a unique model that provides results for a given prediction horizon (h), which renders its implementation difficult. Moreover, prediction information between instants t and t+h is not available.

The parallel structure uses only one model that provides results for several prediction horizons. Indeed, all the predictions from instant t to t+h are obtained in parallel, which renders its execution time interesting.

Nevertheless, after a prediction analysis conducted on these two structures, it proves that they are not suited to our study, as the prediction errors are not acceptable unless the prediction horizon does not exceed the 10 h. It is obviously a too weak horizon to imagine an efficient maintenance of the FC.

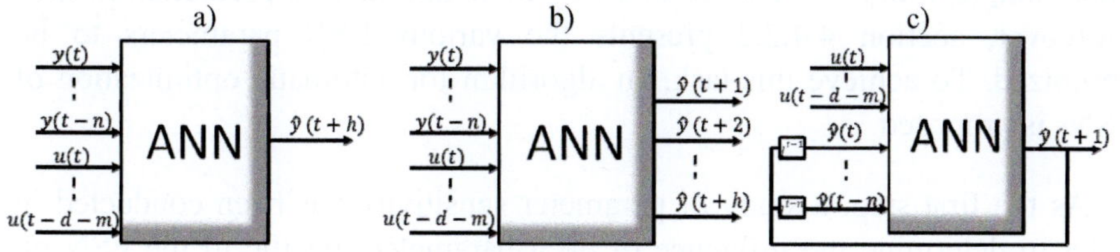

Figure 4.45. *Various structures*

The consideration of wider prediction horizons requires the use of the iterative structure, although it offers a lower precision. Indeed, the use of this structure involves the construction of a prediction model based on a direct structure for the prediction at *t*+1. Using this model, the principle is then to use the achieved prediction as a new input instead of a known point. This action is iterated until obtaining the expected prediction horizon. This method obviously involves propagation of the estimation error, which is itself a consequence of using estimated values instead of real values as inputs. This may prove extremely problematic for each regain in performance stemming from FC characterizations. Indeed, with the use of an iterative structure, it will be impossible for the ESN to know when characterizations take place. Hence, besides the voltage regressors at the model input, there is a Boolean input that passes from state 0 to state 1 when a characterization (U(I) + EIS every 168 h) is added. This makes it possible to easily take into account the regain in performance resulting from the characterization, and above all, not propagating too many significant errors that may cause the model to diverge.

In summary, an iterative structure with voltage regressors as inputs, a Boolean operator informing the characterization moment and the time will be used in the following sections.

Once the prediction structure has been chosen, it is important to properly parameterize the model. Let us recall that ESNs are neural networks with "reservoir" and there are various structures of reservoirs (random reservoir, delay reservoir, simple cycle reservoir, delay reservoir with feedback or simple reservoir with feedback). The choice of this reservoir is obviously important, as it must be chosen in a way to obtain the best prediction results. Moreover, section 4.4.2.2 presents the various ESN parameters to be optimized. To achieve this task, an algorithm for automatic optimization of ESNs is proposed.

As the first step, a study of parameter sensitivity has been conducted in order to determine the influence of each parameter specific to the ESN on the result and also of the choice of the reservoir. The study of parameter sensitivity is conducted with the ANOVA (analysis of variance) method.

As the second step, the optimization of ESN parameters relies on genetic algorithms and then on the use of the Hurst coefficient [HUR 50, MOR 15b, MOR 17].

The ANOVA method, introduced by Taguchi [FOW 95] in the 1950s, is a collection of statistical models and procedures, which enables simultaneous comparisons between several means in order to determine the significant relationships between the variables. In other words, it enables us to obtain the influence of each parameter based on a plan of experience.

A first study on the parameters with the strongest influence on the prediction results shows that the spectral radius of the reservoir matrix $\rho(W_{res})$ has the strongest influence on the result. The spectral radius defines the scale of the weights, meaning the importance attributed to the information that passes between two neurons. Consequently, this parameter cannot be dissociated from the number of neurons in the reservoir. This leads to a significant interaction between the number of neurons and the spectral radius that constitutes the second most influential parameter ($N/\ \rho(W_{res})$). Finally, the number of neurons N comes third. Unsurprisingly, it can be concluded that the choices of spectral radius and of number of neurons are key parameters in obtaining a high-performance model.

Moreover, as previously mentioned, it is possible to have various structures of reservoir. A second ANOVA study integrating this time three parameters ($\rho(W_{res})$, N and reservoir structure) shows that the parameter with the strongest influence is the spectral radius, but the parameter occupying this time the second position is the structure of the reservoir and the number of neurons N is the last.

These results enable us to focus on the optimization of these three parameters. A first optimization relies on genetic algorithms (GA – genetic algorithm) that use various time evolutions of the stack voltage issued from several projects. The GAs define a structure of random reservoir and 50 neurons as the parameters providing the best prediction results. The number of regressors has been included in this study and three regressors should be used. Nevertheless, the choice of spectral radius is more difficult to make, as it varies depending on the studied signal particularly in case of sudden variations of the signal. The method seems efficient for the choice of the reservoir structure and the number of neurons. These two parameters are then set at the values provided above. A trade-off should now be found in order to determine the best spectral radiation regardless of the studied signal.

For this purpose, an innovative study relying on the analogy between the Hurst coefficient and spectral radiation is proposed. The Hurst coefficient is a mathematical quantity that enables the detection of the long-term memory effects of a time series [HUR 50]; hence, it is possible to make a connection with the spectral radiation of the reservoir.

The values of spectral radiation determined with the GAs in the first optimization are resumed depending on the predicted signals. For each studied signal, the Hurst coefficient is calculated and the results obtained show that spectral radiation is connected to the long-term memory of the studied signal. One regression equation defining the spectral radius as a function of Hurst coefficient is then defined. In other words, it is sufficient to calculate the Hurst coefficient of the studied signal in order to determine the spectral radius.

The optimal parameters have been defined, and the next stage involves realizing a long-term prognostics of the evolution of performances of PEMFC. The next sections offer a detailed presentation of the simulation results obtained with the model that has just been established.

4.4.4.3. *Simulation results*

First, the results of training realized on the data of the regional project for a constant load profile are presented. In Figure 4.46, the training is realized on the average of voltages of cell no. 1, 2, 3 and 5 evolving towards 1,700 h. The first 100 h of cell no. 4 are also added in order to explain to the model the first points of the time evolution of the voltage to be predicted.

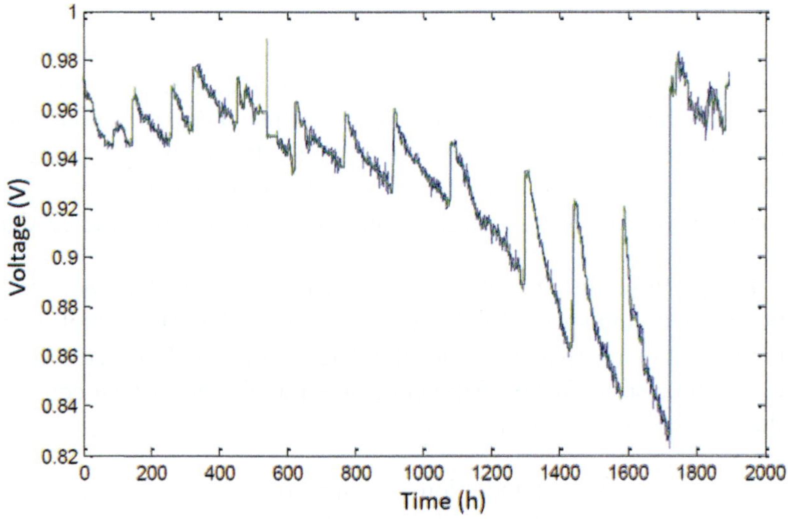

Figure 4.46. *Results of the learning sequence (constant load). For a color version of this figure, see www.iste.co.uk/jemei/cells.zip*

In order to have a good estimation of the performances of the established model, "K-fold" cross-validation method is implemented. It involves splitting the database into k sub-sets. At each iteration, one of the k subsets is used as a test base and the other $k-1$ subsets are concatenated to form the training base.

This method yields the results of the prediction of the evolution of cell no. 4 (Figure 4.47). A purple line is drawn, corresponding to a 10% loss on

PEMFC performances. This threshold value, defined by the "Department of Energy" (DOE) for the PEMFCs intended for transport applications, is considered as a reference in these works in order to define the moment when the PEMFC can no longer respond to the expected profile without a change in the operating conditions.

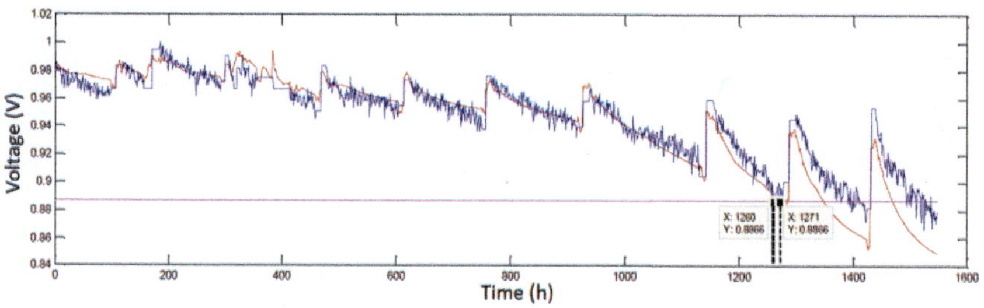

Figure 4.47. *Prediction results (constant load). For a color version of this figure, see www.iste.co.uk/jemei/cells.zip*

The real evolution of the cell voltage is drawn in blue, and the intersection between this plot and the above-defined limit of 10% occurs at t = 1,271 h. This value corresponds to the real RUL. The estimation of the evolution of the cell voltage is drawn in red, and its intersection with the limit of 10% occurs at t = 1,260 h. This value corresponds to the estimated RUL; it is therefore possible to calculate the gap between the real and estimated RUL, which is $\left| \mathrm{RUL} - \widehat{\mathrm{RUL}} \right| = 11\ \mathrm{h}$. The operation is then repeated at regular intervals (in these cases, every 85 h or 5% of the data). The values of real and estimated RUL are plotted in Figure 4.48.

The results presented here are very interesting, which make it possible to obtain an estimated RUL that is quasi-similar to the real RUL. The sought-for purpose is reached. Thanks to this method, a more optimal planning of maintenance operations can thus be established, thereby increasing the lifetime of the FC. It is nevertheless worth noting that the soundest results are only presented here, and they strongly depend on the choice of the training base and on model parameterization.

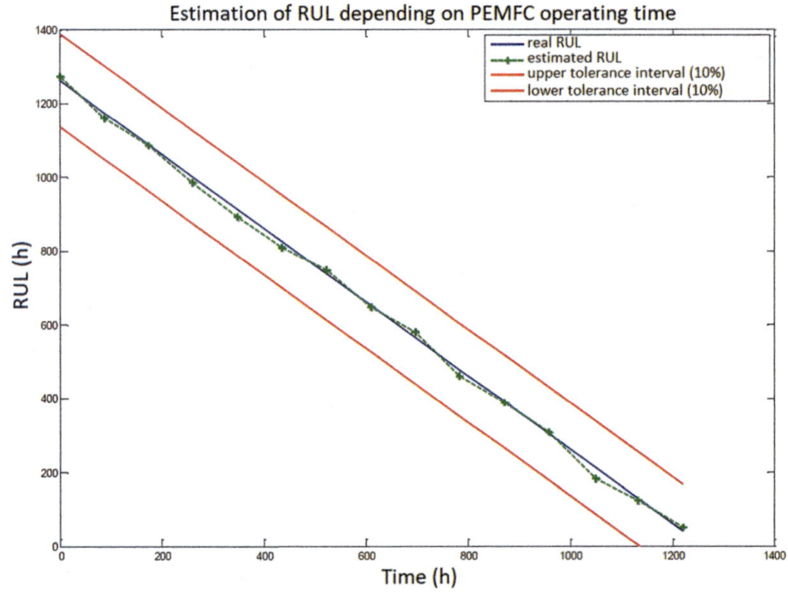

Figure 4.48. *Estimation of RUL throughout the lifetime of the FC (constant load). For a color version of this figure, see www.iste.co.uk/jemei/cells.zip*

Moreover, the results obtained emerge from a constant load stress and the focus is now on the prediction of voltage issued from a dynamic stress.

The data used for training and test are drawn from the SAPPHIRE project. Prediction is realized on cell 3. The model is identical to the one previously used, the only difference being that the current stress is provided as input data. The results are also of high quality (Figure 4.49) with a 3.3% MAPE. A conclusion can thus be drawn on the model capability to predict the evolution of a FC voltage submitted to a dynamic load. It is worth noting that an RUL estimation proves to be complicated, as it is difficult to define the threshold values to implement. This point will be revisited in the conclusion on these works.

Moreover, *a priori* knowledge of the current profile is required, which may be detrimental to certain applications. Furthermore, this method involves the use of a database resulting from experimentation on the lifetime of the FC for the prediction. To cope with this problem, a new prediction algorithm is presented in the following section.

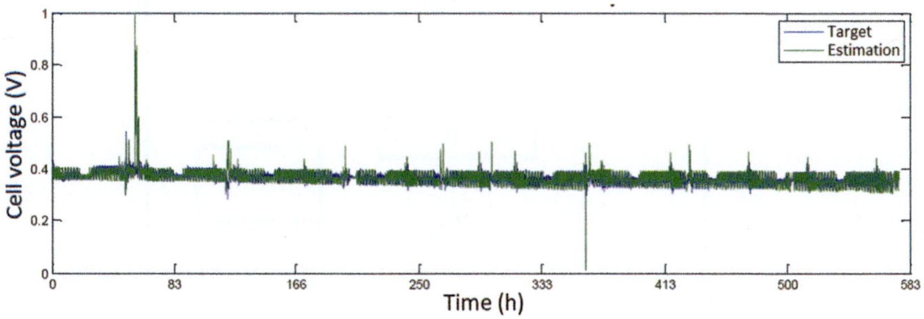

Figure 4.49. *Prediction results (dynamic load). For a color version of this figure, see www.iste.co.uk/jemei/cells.zip*

4.4.4.4. *Further results*

This part focuses on a new prediction algorithm, which is also based on the ESNs, but whose characteristic is to filter the studied signals, thanks to the Hurst coefficient and the wavelet transform. No signal has been filtered for the prediction until present, in order to be as independent as possible from the testing means implemented.

This new algorithm is executed in two stages (Figure 4.50). The first stage uses a combination of wavelets and the Hurst coefficient, and the second stage makes it possible to realize the prognostics.

The algorithm enables the decomposition of the signal into an approximation part and a detail part.

This decomposition is correlated to the Hurst coefficient, which is calculated depending on the studied signal. Let us recall that for a Hurst coefficient $H < 1$, the signal presents strong dynamic variations. As long as H is not greater than or equal to 1, wavelet decomposition is repeated. Then, two ESNs are used in order to predict the approximation part and the detail part(s). The sum of these two signals provides the time signal of the predicted voltage.

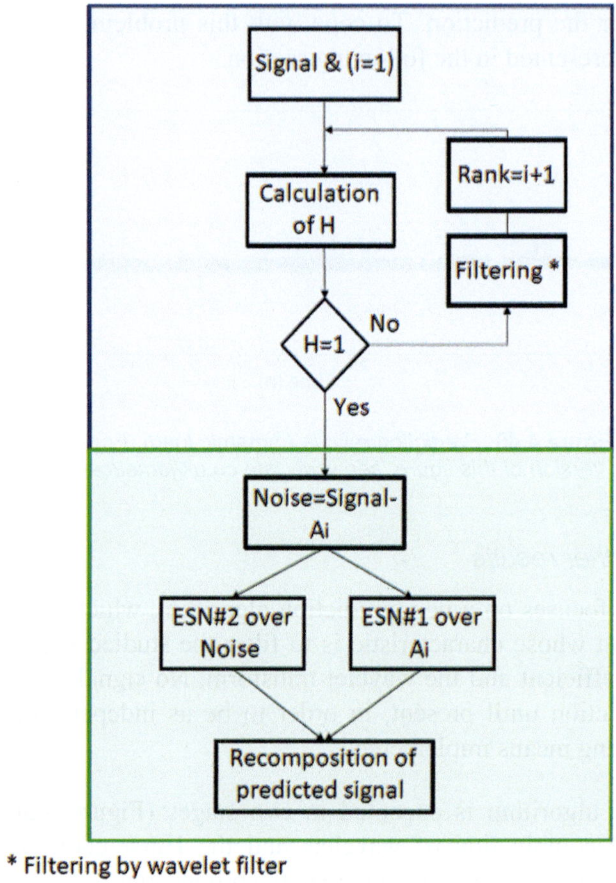

Figure 4.50. *New algorithm proposed for prognostics. For a color version of this figure, see www.iste.co.uk/jemei/cells.zip*

The results are presented in Figure 4.51. The error obtained is low, with an MAPE of 0.97% on the complete recomposed signal. These results are interesting, enabling a correct evaluation of the ageing of the PEMFC on a dynamic load profile in which natural ageing is only taken into account. The results obtained by this method are better than those previously used. It is worth noting that the algorithm proposed here only requires the first 340 h of the signal, which means a five times reduction of the data required for training.

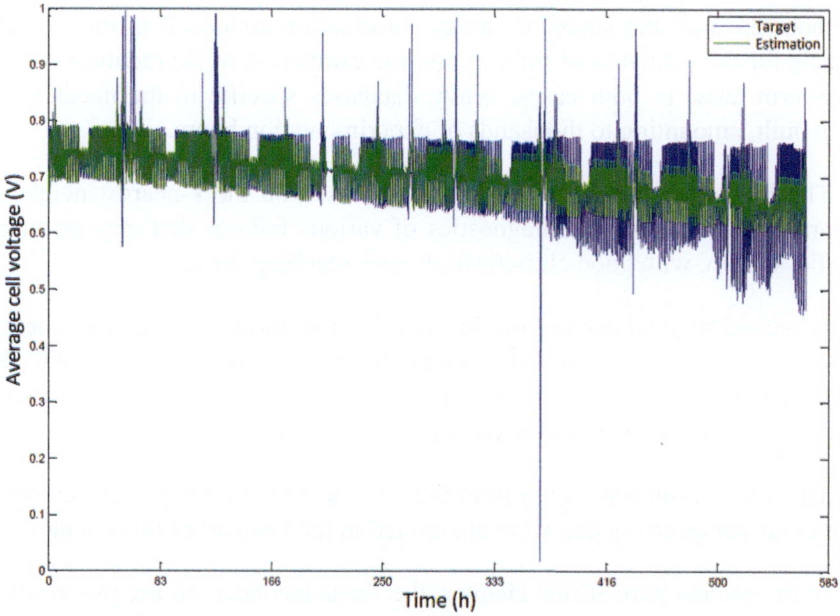

Figure 4.51. *Prediction results with wavelet-based algorithm. For a color version of this figure, see www.iste.co.uk/jemei/cells.zip*

4.5. Conclusion

This last chapter has opened various paths that will eventually lead to more reliable FC systems, whose lifetime meets the industrial expectations. Many works have been summarized in this synthesis, from the study of the degradations of a FC and its system to lifetime prognostics going through its diagnostics.

Indeed, the first part has focused on the study of degradation mechanisms implemented in a FC and the importance of controlling them in order to establish a proper diagnostics following failure or a proper prediction of its remaining useful life.

Thus, the thesis by Elodie Pahon and the ANR DIAPASON2 project have especially made it possible to develop two data-based methods for the diagnostics of PEMFCs. The first method relates to controlled classification of data extracted from electrochemical impedance spectra, for the study of failures in the fuel cell system and for the estimation of ageing. The other method relies on the study of energy information contained in the voltage signals for the detection of failures and the estimation of degradation during long-term tests. In both cases, heavy databases specific to the needs have been built, amounting to thousands of experimentation hours.

The implementation of a methodology based on the k-nearest neighbor algorithm has enabled the diagnostics of various failures that may occur in the FC system, with good classification rates reaching 90%.

A second method relying on the wavelet transform coupled with energy indicators has been developed to estimate the state of health of a FC. In particular, it has made it possible to distinguish between various states of health depending on cathode or cooling circuit failure.

The set of results emerging from these works are of high quality and open numerous perspectives that were elaborated in the last part of this chapter.

In the second part of this chapter, the focus has been on the prognostics of FC systems. For this purpose, the works involved in the theses by Rosa-Elvira Sanchez and Simon Morando have been referred to, as well as the ANR PROPICE regional project and SAPPHIRE European project.

Thus, an initial model predicting the time evolution behavior of the voltage was developed using the ANFIS model, which offers the advantage of performing without an analytical model. This methodology based on the capacity to learn the behavior of the system offers good prediction performances with acceptable calculation times for real-time application. Nevertheless, the training bases prove to be substantial.

In fact, in an attempt to reduce the amount of data involved by model construction and to obtain even higher-performance results, further methodologies have been studied in the thesis by Simon Morando. They also rely on the neural networks, but “with reservoir”. These are specific neural networks whose key characteristic is the use of a dynamic reservoir of neurons. The training algorithm is relatively simple because the weights at reservoir output are only optimized.

Various methods coupling ESNs, genetic algorithms or even the wavelet transform have been developed. The results obtained to date are most promising, enabling a prediction of cell voltages until the end of the FC's lifetime with an error below 10%. Furthermore, the first 200 operating hours of the FC must only be known in order to predict the cell voltages by the end of lifetime. Finally, an estimation of the remaining useful life (RUL) could be displayed, which is very likely a breakthrough in the ESN-based PHM of the FCs.

Summary and Conclusion

Today, terms such as energy transition and global warming have become part of our daily vocabulary. Researchers in the field of energy are aware of the objectives of the "Horizon 2020" program. Enduring solutions should be proposed for the short- and medium-terms. In fact, it has been shown that hydrogen is an option in the energy mix of the future, particularly in relation to fuel cells.

The development of the works described in this book has been centered on the fuel cell. The main technology studied is of PEMFC (proton exchange membrane fuel cell) type, as it offers interesting perspectives for transport and stationary applications transport. Nevertheless, many technological and scientific areas remain to be explored in order to reach an industrial development level. These challenges have been described throughout this book, and above all, solutions have been proposed in order to render this technology more reliable in the short-term. Three main axes have been addressed. The first relates to the fuel cell system and its experimental implementation (Chapter 2). The second axis is dedicated to the hybridization of fuel cell generators (Chapter 3). The last axis focuses on the diagnostics and prognostics of the FC (Chapter 4).

Chapter 1 provided an overview of the current energy situation and highlighted the obsolescence of our energy model. In 2010, 80% of the global primary energy was produced using fossil energies, driving the CO_2 emissions to 30 Gt. Various solutions are obviously available for the decarbonization of the current energy mix. Among them is the hydrogen vector, which when coupled with a fuel cell offers many advantages, regardless of the concerned application. In a transport application, the fuel cell has significantly higher energy efficiency than internal combustion engines or conventional energy conversion systems, and it eliminates polluting emissions and is silent. The drawbacks are its prohibitive cost for a wide-scale application (1,000 $/kW), its very short lifetime and the very low overall efficiency of the system when the fuel cell is considered in its environment.

Some of these drawbacks can nevertheless be solved in the short- or medium-term, depending on their degree of difficulty. For example, the cost is significantly reduced when a transition is made towards mass manufacturing of fuel cells. For this purpose, its lifetime should increase so that it reaches over 5,000 h for transport applications and 80,000 h for stationary applications. However, this challenge is far from being impossible to overcome, as the FCs have no rotating parts and should eventually have very long lifetimes if the preventive and palliative maintenance operations are correctly programed.

The fuel cell is, however, a complex electrochemical generator that is multiphysical (electrical, fluidic, electrochemical, thermal, mechanical, etc.) and multiscale (of time and space). A good understanding of its operation is crucial for its implementation, which requires a surrounding system. Therefore, Chapter 2 presented various FC technologies, with a focus on the PEMFC and SOFC (solid oxide fuel cell). Then, a system approach was proposed for the two types of FC. Various components of this system were highlighted and many interacting phenomena of various natures were presented. The components of the FC system, notably the air compressor, should be optimized in order to increase the overall efficiency and ensure better system dynamics. Finally, a further important part of this chapter was dedicated to the testing means implemented by our research team. Two test benches that were fully designed, developed and instrumented by our team were presented. They enable, among other functions, detailed characterizations in the static or dynamic state of the studied FCs, which are required for the understanding of multiphysical and multiscale phenomena. It was also through these test benches that we constituted many databases required for the establishment of behavioral models or the validation of analytical models. Furthermore, many models, from a single generator to a complete system across the components constituting the system, were developed. They have enabled us to respond to various needs, notably to the energy optimization of the overall system by developing close control laws. Finally, the development of all the data-based methodologies for the diagnostics and prognostics presented in Chapter 4 would have been impossible without these test benches. Overall, the works presented in this chapter were based on the studies conducted on the fuel cell and its system.

Chapter 3 dealt with the hybridization of energy sources. We deliberately focused on transport applications in which dynamic constraints can be the sources of degradation for the FC stack. Two case studies have been proposed, and in each case, the development and the implementation of an intelligent energy management strategy are at the core of our preoccupations.

In the works conducted on the heavy ECCE vehicle composed of a FC, batteries and ultracapacitors, mathematical approaches and artificial intelligence tools have been studied and implemented. They include the wavelet transform, ARIMA (AutoRegressive Integrated Moving Average) models and NARNN (Nonlinear AutoRegressive Neural Network) based on neural networks. Energy management aims to distribute the vehicle power demand to various sources of energy depending on their characteristics and particularly on their frequency range. This ensures the optimal operation of each source throughout the mission and whatever may be the signal dynamics. Thanks to the wavelet transform, the various frequency ranges composing the power demand signal have been identified and sent to the corresponding source (low frequencies to the FC, medium frequencies to the batteries and high frequencies to the ultracapacitors). Nevertheless, this method requires *a priori* knowledge of the power signal, which is not possible for a real-time application. To solve this problem, time series modeling and prediction have been used. Two algorithms have been developed and compared. The adaptive ARIMA model yields interesting results in terms of prediction errors and execution time for a relatively long prediction horizon. As for the NARNN adaptive model, it makes it possible to have a shorter execution time with excellent results for short-term predictions. Nevertheless, these two algorithms provide very good results using only current and previous data of a univariate signal of the power demand of a hybrid vehicle. Furthermore, the use of these algorithms makes it possible to expand the lifetime of the components of the hybrid system and particularly the fuel cell generator. Indeed, these high-performance algorithms enable the preservation of the FC by stressing it in the “proper” operation areas.

In the second case study, energy management for a hybrid locomotive has been implemented. It is composed of batteries, ultracapacitors and an internal combustion engine. The proposed strategy is high performance, real-time, robust and adaptive, which does not require *a priori* knowledge of the mission profile and also takes into account the physical characteristics of the components and embedded sources. This has been obtained by the implementation of a type-2 fuzzy controller whose parameters have been optimized by a genetic algorithm. This study is the first of its kind, which proposes type-2 fuzzy logic optimized by the genetic algorithm for hybrid electric vehicles. Apart from the fact that the specifications imposed by the manufacturer in terms of energy consumption and level of state of charge of batteries and ultracapacitors are respected, the lifetime of the batteries has also been increased due to artificial intelligence tools.

Thus, the methods developed in this chapter on various applications show that it is possible to increase the lifetime of electrochemical generators when they are hybridized thanks to mathematical tools and other tools resulting from artificial intelligence. Chapter 4 proposed the use of these tools for the development of diagnostic and prognostic tools for the FC.

In the last chapter of this book, we dealt with two challenges, namely the reliability and the lifetime of FCs. Very good results have been obtained in relation to the diagnostics, since tools have been implemented and successfully tested on real systems. Referring to the prognostics of the FC, we have laid the foundation of a completely new discipline, which is worth being explored. The first works show that the methods developed offer promising perspectives in terms of dependability.

As an initial step, FC degradation mechanisms have been presented before dealing with the failures that may occur on the FC system. This study has enabled us to evaluate the impact of system failures on the loss of stack performance. Our test benches have enabled us to reproduce a certain number of failures, such as air starvation, cooling and short circuit, and to evaluate their impacts on the performance of the FC. Here the main purpose is to build databases coupled with classification algorithms in order to diagnose failure. Various descriptors have been identified on the EIS, acquired before or after the failure occurrence, in view of the best possible representation of the state of the system. A supervised classification algorithm (k-nearest neighbors) associated with a cross-validation method (leave-one-out) was used to obtain a good classification at a level of 85–90%. In a further study based on the same methodology, we were able to estimate the degradation of the FC with 80% reliability. Finally, the estimation of the state of health has been approached by means of the wavelet transform coupled with energy indicators. The first encouraging results offer interesting perspectives, which will be discussed further on. It is worth mentioning that these algorithms are obviously compatible with a real-time application. In fact, this has been accomplished within the framework of the ANR DIAPASON2 project. The diagnostics of FCs is an important process from the perspective of the maintainability of future systems. In fact, fast detection and precise localization of the failure that leads to FC degradation before the complete stop of the system are necessary. While diagnostics enable increased operation reliability, prognostics should furthermore enable lifetime extension.

The first study on prognostics enabled us to design and realize a model for predicting the degradation of a FC system of PEMFC type. Using a data-oriented approach, we have shown that the measurement of FC voltage enables the prediction of the future behavior of the system. The architecture of the implemented prediction system relies on the Adaptive Neuro-Fuzzy Inference System (ANFIS), which is a smooth and adaptive approach. More precisely, such an approach does not require physical modeling, which is therefore easy to implement. This method also has the capacity to learn the behavior of the system, and ANFIS has good prediction performances, while reducing the training time. Furthermore, due to its short calculation time, this method is well adapted to real-time applications. Nevertheless, even though this first approach has revealed promising tendencies, it has not been possible to define the remaining useful lifetime (RUL).

Therefore, we have developed a methodology centered on a new neural network system called the echo state network, in which algorithm complexity is replaced by structural complexity, so that the training phase is more rapid than that of traditional neural networks. However, the construction of an optimized ESN is still a task requiring the user to define the structure and parameters of the ESN depending on the problem under study. An innovative study has been proposed in order to automatically optimize ESN parameters. It relies on the use of genetic algorithms and the Hurst coefficient. Thanks to this method, voltage evolution is predicted throughout the life of the FC with only some training data. Moreover, the RUL estimated at each instant of the life of the FC shows an error below 10% compared with the real RUL. Hence, a more optimal planning of maintenance operations could be established, which would enable an extension of the FC's lifetime. This study also shows that the use of ESNs as a prediction tool is reliable, and to our knowledge, it is a worldwide premiere in the field of ESNs applied to PHM of the FC.

Summarizing this last part of our works concerning prognostics, it is worth noting that the developments made should enable us in the short-term to use the raw data acquired on the real FC systems to define an indicator of RUL. This objective has been partly reached. Nevertheless, the reliability of this estimation should be increased, as for the moment it is highly dependent on the choice of the training base. This also enables the development of a generic approach, which will certainly be data oriented. In the case of FCs where there is a lack of understanding of the involved phenomena, data-oriented prognostics seems to be the most reliable approach.

Furthermore, let us recall that all these developments could not have been achieved without our experimental means. Although the development of test benches does not seem to be a scientific subject in its own right, the added value that this activity brings to our research efforts is obvious. We must acknowledge the excellent work conducted by our engineers, PhD students and postdoctoral students on the test benches for more than 15 years within the research federation CNRS FCLAB and at FEMTO-ST Institute.

References

[ADD 02] ADDISON P.S., *The Illustrated Wavelet Transform Handbook: Introductory Theory and Applications in Science, Engineering, Medicine and Finance*, Institute of Physics Publishing, London, 2002.

[ADE 16] ADEME, Gestion des réseaux par l'injection d'hydrogène pour décarboner les énergies: vecteur hydrogène, available at: www.ademe.fr/invest-avenir, 2016.

[AFH 18] AFHYPAC, Production d'hydrogène par électrolyse de l'eau, Memento de l'hydrogène fiche 3.2.1, available at: http://www.afhypac.org/documents/tout-savoir/Fiche%203.2.1%20-%20Electrolyse%20de%20l%27eau%20rev%20%20Janv.%202018-2%20ThA.pdf, 2018.

[AFN 88] AFNOR, Maintenance industrielle, Recueil de normes françaises, 1988.

[AGB 11] AGBLI K., PÉRA M., HISSEL D. *et al.*, "Multiphysics simulation of a PEM electrolyser: Energetic macroscopic representation approach", *International Journal of Hydrogen Energy*, vol. 36, no. 2, pp. 1382–1398, 2011.

[AKL 08] AKLI C.A., Conception systèmique d'une locomotive hybride de démonstration et d'investigations en énergetique LHyDIE développée par la SNCF, PhD thesis, Institut National Polytechnique de Toulouse, 2008.

[ALL 10] ALLÈGRE A.L., Méthodologies de modélisation et de gestion de l'énergie de stockage mixtes pour véhicules électriques et hybrides, PhD thesis, Université de Lille 1, 2010.

[ALL 11] ALLAF O., KADER S.A., "Nonlinear autoregressive neural network model for estimation soil temperature: A comparision of different optimization neural networl algorithms", *IEEE International Conference on Industrial Technology*, Auburn, USA, vol. 143, 2011.

[ANR 10] ANR, "Investigations and characterization and development of compressor for FC systems with powers above 10 kW - ICARE CSP", available at: http://www.agence-nationale-recherche.fr/Projet-ANR-08-PANH-0010, 2010.

[ANR 14] ANR, Project DIAPASON2, available at: http://www.agence-nationalerecherche.fr/?Projet=ANR-10-HPAC-0002, 2011–2014.

[AYA 04] AYAD M-Y., Mise en oeuvre des supercondensateurs dans les sources hybrides Continues, PhD thesis, electrical engineering, Institut national polytechnique de Lorraine, available at: http://www.green.uhp-nancy.fr, 2004.

[BAD 13] BADIN F. *et al.*, *Le stockage de l'énergie*, edited by ODRU P., collection UniverSciences, Dunod, Paris, 2013.

[BAE 12a] BAERT J., JEMEÏ S., CHAMAGNE D. *et al.*, "Energetic macroscopic representation of a naturally-aspirated engine coupled to a salient pole synchronous machine", *IFAC-PPPSC 2012*, Toulouse, France, pp. 1–6, 2012.

[BAE 12b] BAERT J., JEMEÏ S., CHAMAGNE D. *et al.*, "Modeling and energy management strategies of a hybrid electric locomotive", *IEEE Vehicle Power and Propulsion Conference (VPPC)*, Seoul, South Korea, 9–12 October 2012.

[BAE 13a] BAERT J., Hybrid electric locomotives: contributions to modelling and type-2 fuzzy logic energy management Strategy, PhD thesis, Université de Franche-Comté, 2013.

[BAE 13b] BAERT J., Gestion d'énergie pour locomotive électrique hybride basée sur la logique floue d'ordre 2 par approche génétique évolutionnaire, *JCGE*, 2013.

[BAE 14] BAERT J., JEMEÏ S., HISSEL D. *et al.*, "Energetic macroscopic representation and optimal fuzzy logic energy characterization of nickel-cadmium battery cells", *EPE Journal*, 2014.

[BAR 13] BARBIR F., *PEM Fuel Cells, Theory and Practice*, 2nd edition, Elsevier Ltd, 2013.

[BEN 94] BENGIO Y., SIMARD P., FRASCONI P., "Learning long-term dependencies with gradient is difficult", *IEEE Transaction on Neural Network*, vol. 5, pp. 157–166, 1994.

[BEN 04] BENTOUMI M., Outils pour la détection et la classification Application au diagnostic de défauts de surface de rail, PhD thesis, Université Henri Poincaré, Nancy 1, 2004.

[BEN 05] BEN AHMED H., MULTON B., ROBIN G. *et al.*, "Le stockage de l'énergie dans les applications stationnaires", *Revue technologie - Sciences et techniques industrielles*, no. 136, pp. 60–66, March 2005.

[BEN 14a] BENOUIOUA D., CANDUSSO D., HAREL F. *et al.*, "PEMFC stack voltage singularity measurement and fault classification", *International Journal of Hydrogen Energy*, vol. 39, no. 36, pp. 21631–21637, 2014.

[BEN 14b] BENOUIOUA D., CANDUSSO D., HAREL F. *et al.*, "Fuel cell diagnosis method based on multifractal analysis of stack voltage signal", *International Journal of Hydrogen Energy*, vol. 39, no. 5, pp. 2236–2245, 2014.

[BER 14] BERNIER J.C. *et al.*, "Chimie et enjeux énergétique", *EDP Sciences*, 2014.

[BOU 07] BOUDELLAL M., *La pile à combustible: structure, fonctionnement, applications*, Dunod, Paris, 2007.

[BOU 09] BOULON L., Modélisation multiphysique des éléments de stockage et de conversion d'énergie pour les véhicules électriques hybrides. Approche systémique pour la gestion d'énergie, PhD thesis, Université de Franche-Comté, 2009.

[BOX 08] BOX G., JENKINS G., REINSEL G., *Time Series Analysis: Forecasting and Control*, John Wiley and Sons Ltd, 2008.

[BRE 15] BRESSEL M., HILAIRET M., HISSEL D. *et al.*, "Dynamical modeling of Proton Exchange Membrane Fuel Cell and parameters identification", *6th International Conference on Fundamentals & Development of Fuel Cells (FDFC)*, 2015.

[BRO 02] BRODRICK C.J., LIPMAN T., FARSHCHI M. *et al.*, "Evaluation of fuel cell auxiliary power units for heavy-duty diesel trucks", *Journal of Transportation Research, part D7*, pp. 303–315, 2002.

[BRY 11] BRYANT A., MAWBY P., TAVNER P., "An industry-based survey of reliability in power electronic converters", *IEEE Transactions on Industry Applications*, vol. 47, no. 3, pp. 1441–1451, May 2011.

[CAD 14] CADET C., JEMEÏ S., DRUART F. *et al.*, "Diagnostic tools for PEMFCs: From conception to implementation", *International Journal of Hydrogen Energy*, vol. 39, no. 20, pp. 10613–10626, 3 July 2014.

[CAN 07] CANDUSSO D., GLISES R., HISSEL D. *et al.*, "Pile à combustible PEMFC et SOFC. Description et gestion du système", *Techniques de l'Ingénieur*, 2007.

[CAN 13] CANDUSSO D., Contribution à l'expérimentation de générateurs à piles à combustible de type PEM pour les systèmes de transport, HDR, École Normale Supérieure de Cachan, 2013.

[CAR 08] CARTER R., CRUDEN A., "Strategies for control of a battery/supercapacitor system in an electric vehicle", *IEEE, SPEEDAM'08*, Capri, Italy, June 2008.

[CAU 10] CAUX S., HANKACHE W., FADEL M. *et al.*, "On-line fuzzy energy management for hybrid fuel cell systems", *International Journal of Hydrogen Energy*, vol. 35, no. 5, pp. 2134–2143, 2010.

[CEA 07] CEA, Pile à combustible, no. 87, available at: www.cea-technologies.com/articles/article/627/fr, 2007.

[CEA 18] CEA, Liten, available at: http://liten.cea.fr/cea-tech/liten, 2018.

[CHA 07] CHANG P.-C., CHEN S-S., KO Q.H. *et al.*, "A genetic algorithm with injecting artificial chromosomes for single machine scheduling problems", *IEEE Symposium on Computational Intelligence in Scheduling, 2007. SCIS'07*, pp. 1–6, April 2007.

[CHA 10] CHAN C.C., BOUSCAYROL A., CHEN K., "Electric, hybrid, and fuel-cell vehicles: Architectures and modeling", *IEEE Transactions on Vehicular Technology*, vol. 59, no. 2, pp. 589–598, 2010.

[CHA 13] CHATTI N., OULD-BOUAMAMA B., GEHIN A.-L. *et al.*, "Merging bond graph and signed directed graph to improve FDI procedure", *European Control Conference (ECC)*, pp. 1457–1462, 2013.

[CHE 06] CHENDEB M., KHALIL M., DUCHÊNE J., "Methodology of wavelet packet selection for event detection", *Signal Processing*, vol. 86, pp. 3826–3841, 2006.

[CHU 13] CHUJAI P., KERDPROSOP N., KERDPROSOP K., "Time series analysis of houshold electric consumption with ARIMA and ARMA models", in *Proceedings of the International Multi Conference of Engineers and Computer Scientists IMECS 2013*, Hong Kong, vol. 1, 2013.

[ÇÖG 15] ÇÖGENLI M.S., MUKERJEE S., YURTCAN A.B., *Membrane Electrode Assembly with Ultra Low Platinum Loading for Cathode Electrode of PEM Fuel Cell by Using Sputter Deposition, Fuel Cells*, Wiley Online Library, 2015.

[COL 06] COLLIER A., WANG H., ZIYUAN X. *et al.*, "Degradation of polymer electrolyte membranes", *International Journal Hydrogen Energy*, vol. 31, no. 13, pp. 1838–1854, October 2006.

[COM 03] COMMISSION EUROPÉENNE, Hydrogène et Piles à combustible - Une vision pour notre avenir, Rapport final du groupe de haut niveau - Office des publications officielles des Communautés Européennes, EUR 20719FR, ISBN 92-894-6284-1, Luxembourg, 2003.

[DAV 10] DAVAL L., Review on Compressor for FC Systems with Powers above 10 kW, Master Degree Report, University of Franche-Comté, 2010.

[DEV 14a] DEVILLERS N., JEMEÏ S., PÉRA M.C. *et al.*, "Review of characterization methods for supercapacitor modeling", *Journal of Power Sources*, vol. 246, pp. 596–608, 15 January 2014.

[DEV 14b] DEVILLERS N., PERA M.C., JEMEÏ S. *et al.*, "Complementary characterization methods for lithium-Ion polymer secondary battery modeling", *International Journal of Electrical Power & Energy Systems*, vol. 67, pp. 168–178, 2014.

[DOU 05] DOUGLAS H., PILLAY P., "The impact of wavelet selection on transient motor current signature analysis", *IEEE International Conference on Electric Machines and Drives*, San Antonio, 2005.

[DRE 02] DREYFUS G., MARTINEZ J.M., SAMUELIDES M., *Réseaux de neurones: méthodologie et applications*, Eyrolles, 2002.

[DUB 90] DUBUISSON B., *Diagnostic et Reconnaissance des Formes*, Hermès, Paris, 1990.

[ESP 06] ESPANET C., KAUFFMANN J., BERNARD R., "Comparison of two in-wheel permanent magnent motors for military applications", *Vehicle Power and Propulsion Conference VPPC'06, IEEE*, pp. 1–6, 2006.

[EXP 14] EXPLICIT, Schéma directeur pour le développement des énergies issues de sources renouvelables et des déchets, Report, 2014.

[FER 08] FERREIRA A., POMILIO J.A., SPIAZZI G. *et al.*, "Energy management fuzzy logic supervisory for electric vehicle power supplies system", *IEEE, Transaction on Power Electronics*, vol. 23, no. 1, pp. 107–115, January 2008.

[FER 13] FERHOUNE M., Caractérisation des batteries Ni-Cd et vieillissement accéléré pour une application ferroviaire, Master's thesis, Université de Franche-Comté, 2013.

[FOW 95] FOWLKES W.Y., CREVELING C.M., *Engineering Methods for Robust Product Design: Using Taguchi Methods in Technology and Product Development Reading*, Addison-Wesley, MA, 1995.

[FRA 15] FRANCE P.E., MATEO P., *Hydrogène: la transition énergétique en marche!*, Gallimard, 2015.

[FRE 13] FRERIS L., *Les énergies renouvelables pour la production d'électricité*, Dunod, Paris, 2013.

[FRF 14a] FR FCLAB, IEEE PHM data challenge 2014, available at: http://eng.fclab.fr/ieee-phm-2014-data-challenge/, 2014.

[FRF 14b] FR FCLAB, IEEE Prognostic and Health Management data challenge 2014, available at: http://eng.fclab.fr/ieee-phm-2014-data-challenge/, 2014.

[FRI 03] FRIEDE W., Modélisation et caractérisation d'une pile à combustible du type PEM, PhD thesis, Institut national polytechnique de Lorraine, 2003.

[FU 74] FU K.S., *Syntactic Methods in Pattern Recognition*, Academic Press, Inc, New York, 1974.

[GAI 06] GAINES L., SANTINI D., Economic Analysis of Commercial Idling Reduction Technologies, Argonne National Laboratory, available at: http://www.transportation.anl.gov/pdfs/TA/372.pdf, 2006.

[GAO 01] GAO X., OVASKA S., "Soft computing methods in motor fault diagnosis", *Applied Soft Computing*, vol. 1, no. 1, pp. 73–81, June 2001.

[GAO 11] GAO R., YAN R., *Wavelet, Theory and Application of Manufacturing*, Springer US, 2011.

[GAY 10] GAY C., HISSEL D., LANZETTA F. *et al.*, "Energetic macroscopic representation (EMR) of a solid oxide fuel cell (SOFC) for stirling engine combined cycle in high-efficient powertrains", *IEEE VPPC'2010: Vehicle Power and Propulsion*, Lille, France, September 1–3, 2010.

[GOU 15] GOURIVEAU R., Contribution à l'optimisation des processus de prédiction et de classification pour le Prognostics and Health Management; HDR, 2015.

[GRO 39] GROVE W.R., "On voltaic series and the combination of gases by platinum", *The London and Edinburgh Philosophical Magazine and Journal of Science*, vol. 14, no. 86, pp. 127–130, United Kingdom, 1839.

[GUI 09] GUIDI G., UNDELAND T.M., HORI Y., Effectiveness of supercapacitors as power-assist in pure EV using sodium-nickel chloride battery as main energy storage, EVS24, Stavanger, Norway, May 2009.

[HAJ 06] HAJIMIRI M., SALMASI F., "A fuzzy energy management strategy for series hybrid electric vehicle with predictive control and durability extension of the battery", *IEEE Conference on Electric and Hybrid Vehicles, 2006. ICEHV'06*, pp. 1–5, 2006.

[HAM 08] HAMOU MAMAR Z., Analyse temps-échelle et reconnaissance des formes pour le diagnostic du système de guidage d'un tramway sur pneumatiques, PhD thesis, Université Blaise Pascal – Clermont II, July 2008.

[HAM 14] HAMAZ T., Outils de caractérisation et de diagnostic d'une pile à combustible de type PEM par mesure du champ électromagnétique externe, Thesis, Université de Grenoble, November 2014.

[HES 08] HESS A., STECKI J.S., RUDOV-CLARK S.D., "The maintenance aware design environment: Development of an aerospace PHM software tool", *Proceedings PHM08*, 2008.

[HIS 04] HISSEL D., Modélisation, optimisation énergétique et diagnostic de systèmes piles à combustible, HDR, Université de Franche-Comté, 2004.

[HIS 06] HISSEL D., KAUFFMANN J.M., "Fuel cells and their applications in Belfort (France)", *Fuel Cells*, vol. 6, no. 1, pp. 3–3, 2006.

[HIS 07] HISSEL D., CANDUSSO D., HAREL F., "Fuzzy-clustering durability diagnosis of polymer electrolyte fuel cells dedicated to transportation applications", *IEEE Transaction on Vehicular Technology*, vol. 56, no. 5, pp. 2414–2420, 2007.

[HOC 14] HOCHSTEIN A., AHN H.I., LEUNG Y.T. *et al.*, "Switching vector autoregressive models with higher-order regime dynamics", *IEEE PHM Conference 2014*, pp. 1–10, June 2014.

[HOO 03] HOOGERS G., "Automotive applications", in HOOGERS G. (ed.), *Fuel Cell Technology Handbook*, CRC Press, Boca Raton, 2003.

[HUR 50] HURST H.E., DIVISION A.S.O.C.E.H., "Long-term storage capacity of reservoirs", *American Society of Civil Engineers*, 1950.

[IBR 13] IBRAHIM M., Algorithmes de gestion en ligne des flux énergétiques dans les véhicules hybrides électriques, PhD thesis, Université de Franche-Comté, 2013.

[IBR 15a] IBRAHIM M., ANTONI U., YOUSFI STEINER N. *et al.*, "Signal-based diagnostics by wavelet transform for proton exchange membrane fuel cell", *Energy Procedia*, vol. 74, pp. 1508–1516, 2015.

[IBR 15b] IBRAHIM M., JEMEÏ S., YOUSFI STEINER N. *et al.*, "Selection of mother wavelet and decomposition level for energy management in electrical vehicles including a fuel cell", *International Journal of Hydrogen Energy*, vol. 40, no. 45, pp. 15823–15833, June 2015.

[IBR 16] IBRAHIM M., JEMEÏ S., WIMMER G. *et al.*, "Nonlinear autoregressive neural network in an energy management strategy for battery/ultra-capacitor hybrid electrical vehicles", *Electric Power Systems Research*, vol. 136, pp. 262–269, 2016.

[INT 11] INTERNATIONAL ENERGY AGENCY, World Energy Outlook 2011, available at: http://www.iea.org/publications/freepublications/publication/weo-2011.html, 2011.

[INT 12] INTERNATIONAL ENERGY AGENCY, World Energy Outlook 2012, available at: http://www.iea.org/publications/freepublications/publication/world-energy-outlook-2012.html, 2012.

[INT 15] INTERNATIONAL ENERGY AGENCY, Energy and Climate Change, World Energy Outlook Special Report, available at: https://www.iea.org/publications/freepublications/publication/weo-2015-special-report-energy-climate-change.html, 2015.

[ISA 12] ISAR A., "Space-frequency localization as bivariate mother wavelets selecting criterion for hyperanamytic bayesian image denoing", *Fluctuations and Noise Letters*, vol. 11, 2012.

[ISM 05] ISMAIL Z., MAHPOL K., "SARIMA model for forecasting malaysian electricity generated", *IEEE Transactions on Power Systems*, vol. 21, pp. 143–152, 2005.

[ISO 04] ISO13381-1, Condition monitoring and diagnostics of machines prognostics Part 1: General guidelines, International Standard, ISO, 2004.

[ISO 06] ISO13374-2, Condition monitoring and diagnostics of machines – data processing, communication and presentation – part 2: Data processing, International Standard, ISO, 2006.

[JAE 01] JAEGER H., "The "echo state" approach to analysing and training recurrent neural networks, Technical Report GMD 148", *German National Research Center for Information Technology*, 2001.

[JAE 02] JAEGER H., "Tutorial on training recurrent neural networks, covering BPPT, RTRL, EKF and the echo state network approach, Technical report". *GMD—German National Research Institute for Computer Science*, 2002.

[JAE 03] JAEGER H., "Adaptive nonlinear system identification with Echo State Networks", in BECKER S., THRUN S., OBERMAYER K. (eds), *Advances in Neural Information Processing Systems*, vol. 15, pp. 609–616, MIT Press, 2003.

[JAM 09a] JAMMEH E., FLEURY M., WAGNER C. *et al.*, "Interval type-2 fuzzy logic congestion control for video streaming across IP networks", *IEEE Transactions on Fuzzy Systems*, vol. 17, no. 5, pp. 1123–1142, 2009.

[JAM 09b] JAMALUDIN J., RAHIM N., HEW W., "Development of a selftuning fuzzy logic controller for intelligent control of elevator systems", *Engineering Applications of Artificial Intelligence*, vol. 22, no. 8, pp. 1167–1178, 2009.

[JAM 10] JAMES B., Hydrogen and Fuel Cells Program and Vehicle Technologies Program Annual Merit Review and Peer Evaluation Meeting (AMR), Directed Technologies, Inc. The 2010 U.S. Department of Energy (DOE), Washington DC, 2010.

[JAM 12] JAMES B.D., "Fuel cell transportation cost analysis, preliminary results", *Proceedings of the 2012 US Department of Energy (DOE) Hydrogen and Fuel Cells Program and Vehicle Technologies Program Annual Merit Reviex and Peer Evaluation Meeting*, Arlington, Virginia, available at: www.hydrogen.energy.gov/annual_review12_fuelcells.html#analysis, 14–18, 2012.

[JAN 93] JANG J.-S., "ANFIS: Adaptive-network-based fuzzy inference system", *IEEE Transactions on Systems, Man, and Cybernetics*, vol. 23, no. 3, pp. 665–685, 1993.

[JAV 15] JAVED K., GOURIVEAU R., ZERHOUNI N., “Data-driven prognostics of proton exchange membrane fuel cell stack with constraint based summation-wavelet extreme learning machine”, *6th International Conference on Fundamentals and Development of Fuel Cells, FDFC’15*, Toulouse, France, pp. 1–8, February 2015.

[JEM 04] JEMEÏ S., Modélisation neurale d’une pile à combustible de type PEM, PhD thesis, Université de Franche-Comté, 2004.

[JEM 08a] JEMEÏ S., MULOT J., PERA M.C. *et al.*, “Electric power generation system towards transportation applications: Hybridization of a solid oxide fuel cell auxiliary power unit and batteries”, *Fundamentals and Developments of Fuel Cells, Conference 2008*, Nancy, France. p. 143, December 2008.

[JEM 08b] JEMEÏ S., HISSEL D., PÉRA M.C. *et al.*, “A new modeling approach of embedded fuel cell power generators based on artificial neural network”, *IEEE Transactions on Industrial Electronics*, vol. 55, no. 1, pp. 437–447, 2008.

[JEM 10] JEMEÏ S., MULOT J., PÉRA M.C. *et al.*, “Test and characterization of Delphi’s solid oxide fuel cell auxiliary power units (SOFC APU) fed with natural gas and Diesel fuel”, *9th European Solid Oxide Fuel Cell Forum*, Switzerland, 2–161 - 2–177, 2010.

[JEM 11] JEMEÏ S., ESPANET C., HISSEL D. *et al.*, ICARE-CSP: Investigations, CARactérisations et dEveloppement de Compresseurs pour Systèmes PAC de puissance supérieure à 10 kW, Journées plénières du GDR Piles à Combustible et Systèmes (GDR 339), Nantes, June 2011.

[JIN 06] JINRUI N., FENGCHUN S., QINGLIAN R., “A study of energy management system of electric vehicles”, *IEEE*, VPPC’06, Windsor, UK, September 2006.

[JOH 07] JOHN R.I., COUPLAND S., “Type-2 fuzzy logic: A historical view”, *IEEE Computational Intelligence Magazine*, vol. 2, pp. 57–62, 2007.

[JOU 13] JOUIN M., GOURIVEAU R., HISSEL D. *et al.*, “Prognostics and health management of PEMFC State of the art and remaining challenges”, *International Journal of Hydrogen Energy*, vol. 38, no. 35, pp. 15307–15317, 2013.

[JOU 14a] JOUIN M., GOURIVEAU R., HISSEL D. *et al.*, “Prognostics of PEM fuel cell in a particle filtering framework”, *International Journal of Hydrogen Energy*, vol. 39, no. 1, pp. 481–494, 2014.

[JOU 14b] JOUIN M., GOURIVEAU R., HISSEL D. *et al.*, “Prognostics of proton exchange membrane fuel cell stack in a particle filltering framework including characterization disturbances and voltage recovery”, *Proceedings of the 2014 IEEE International Conference on Prognostics and Health Management*, 2014.

[JOU 15a] JOUIN M., Contribution au pronostic d'une pile à combustible de type PEMFC – approche par filtrage particulaire, PhD thesis, Université de Franche-Comté, December 2015.

[JOU 15b] JOUIN M., GOURIVEAU R., HISSEL D. *et al.*, "Particle filters prognostics for PEMFC power prediction at constant current solicitation", *IEEE Transaction on Reliability*, 2015.

[JOU 16] JOUIN M., BRESSEL M., MORANDO S. *et al.*, "Estimating the end-of-life of PEM fuel cells: Guidelines and metrics", *Applied Energy*, vol. 177, pp. 87–97, 2016.

[KAD 09] KADRI K.E., Contribution à la conception d'un générateur hybride d'énergie électrique pour véhicule: modélisation, simulation, dimentionnement, PhD thesis, Université de Technologie de Belfort- Montbéliard, 2009.

[KAW 08] KAWAMURA A., PAVLOVSKY M., TSURUTA Y., "State-of-the-art high power density and high efficiency DC-DC chopper circuits for HEV and FCEV applications", *13th International Power Electronics and Motion Control Conference*, vol. 1, no. d. IEEE, pp. 7–20, September 2008.

[KAZ 03] KAZLOWSKI J., "Electromechanical cell prognotics using online impedence measurments and model-based data fusion techniques", *Proceedings of the 2003 IEEE Aerospace Conference*, vol. 7, pp. 3257–3270, 2003.

[KEL 05] KELLY M., EGGER B., FAFILEK G. *et al.*, "Conductivity of polymer electrolyte membranes by impedance spectroscopy with microelectrodes", *Solid State Ionics*, vol. 176, nos 25–28, pp. 2111–2114, August 2005.

[KER 09] KERMANI S., Gestion énergétique des véhicules hybrides: de la simulation à la commande temps réel, PhD thesis, Université de Valenciennes et du Hainaut Cambrésis, p. 213, 2009.

[KIM 14] KIM T., KIM H., HA J. *et al.*, "A degenerated equivalent circuit model and hybrid prediction for state-of-health (SOH) of PEM fuel cell", *IEEE PHM Conference 2014*, pp. 1–7, June 2014.

[KIS 10] KISHOR N., MOHANTY S.R., "Fuzzy modeling of fuel cell based on mutual information between variables", *International Journal of Hydrogen Energy*, vol. 35, no. 8, pp. 3620–3631, 2010.

[KOC 12] KOCHA S.S., "Chapter 3 – electrochemical degradation: Electrocatalyst and support durability", in MENCH M.M., KUMBUR E.C., VEZIROGLU T.N. (eds), *Polymer Electrolyte Fuel Cell Degradation*, Academic Press, pp. 89–214, 2012.

[KON 03] KONRAD G., SOMMER M., LOSCHKO B. *et al.*, "System design for vehicle applications: Daimler chrysler", in VIELTSTICH W., LAMM A., GASTEIGER H. (eds), *Handbook of Fuel Cell Technology – Fundamentals, Technology and Applications*, vol. 4, John Wiley & Sons Inc., New York, pp. 693–713, 2003.

[KON 15] KONNO N., MIZUNO S., NAKAJI H. *et al.*, "Development of compact and high-performance fuel cell stack", *SAE International Journal of Alternative Powertrains*, vol. 4, no. 1, pp. 123–129, doi:10.4271/2015-01-1175, 2015.

[KOS 08] KOSEKI H., YOSHIMURA K., "Estimation of electromechanical models of power system using ARMA model based on gross spectrum", *2nd IEEE International Conference on Power and Energy PECon08*, Malaysia, 2008.

[KUL 10] KULKARNI C., BISWAS G., KOUTSOUKOS X. *et al.*, "Integrated diagnostic/prognostic experimental setup for capacitor degradation and health monitoring", *2010 IEEE Autotestcon*, pp. 1–7, September 2010.

[KUN 06] KUNDU S., FOWLER M., SIMON L., "Morphological features (defects) in fuel cell membrane electrode assemblies", *Journal of Power Sources*, vol. 157, no. 2, pp. 650–656, July 2006.

[KUN 12] KUNUSCH C., PULESTON P., MAYOSKY M., *Sliding-Mode Control of PEM Fuel Cells*, Springer Verlag, 2012.

[KUR 08] KURZ T., HAKENJOS A., KRÄMER J. *et al.*, "An impedance-based predictive control strategy for the state-of-health of PEM fuel cell stacks", *Journal of Power Sources*, vol. 180, pp. 742–747, 2008.

[KUR 14] KURIA KIMOTHO J., MEYER T., SEXTRO W., "PEM fuel cell prognostics using particle filter with model parameter adaptation", *IEEE PHM Conference 2014*, pp. 1–6, June 2014.

[KUT 10] KUTTER S., BÄKER B., "Predictive online control for hybrids - resolving the conflict between global optimality, robustness and real-time capability", *IEEE*, vol. 978-1-4244-8218-4/10/, pp. 1–6, 2010.

[LAG 13] LAGHROUCHE S., MATRAJI I., AHMED F.S. *et al.*, "Load governor based on constrained extremum seeking for PEM fuel cell oxygen starvation and compressor surge protection", *International Journal of Hydrogen Energy*, vol. 38, no. 33, pp. 14314–14322, 4 November 2013.

[LAP 87] LAPEDES A., FARBER R., Nonlinear signal processing using neural network: Prediction and modeling, Technical report, LA-UR 87-2662, Los Alamos National Laboratory, New Mexico, 1987.

[LAR 10] LARGER L., DUDLEY J.M., "Nonlinear dynamics optoelectronic chaos", *Nature*, vol. 465, no. 7294, pp. 41–42, 2010

[LEC 15] LECHARTIER E., LAFFLY E., PÉRA M.C. *et al.*, "Proton exchange membrane fuel cell behavioral model suitable for prognostics", *International Journal of Hydrogen Energy*, vol. 40, pp. 8384–8397, 26 July 2015.

[LES 10] LESCOT J., SCIARRETTA A., CHAMAILLARD Y. *et al.*, "On the integration of optimal energy management and thermal management of hybrid electric vehicles", *Vehicle Power and Propulsion Conference (VPPC), 2010 IEEE*, pp. 1–6, September 2010.

[LI 14a] LI Z., Data-driven fault diagnosis for PEMFC systems, PhD thesis, Université de Franche-Comté et d'Aix-Marseille, 2014.

[LI 14b] LI Z., GIURGEA S., OUTBIB R. *et al.*, "Online diagnosis of PEMFC by combining support vector machine and fluidic model", *Fuel Cells*, vol. 14, no. 3, pp. 448–456, 2014.

[LI 16] LI Z., GIURGEA S., OUTBIB R. *et al.*, "Online implementation of SVM based fault diagnosis strategy for PEMFC systems", *Applied Energy*, vol. 164, pp. 284–293, 2016.

[LIA 08] LIAO R., CHAN C., HROMEK J. *et al.*, "Fuzzy logic control for a petroleum separation process", *Engineering Applications of Artificial Intelligence*, vol. 21, no. 6, pp. 835–845, 2008.

[LIN 96] LIN T., HORNE B., GILES C., "How memory order effect the performance of NARX networks", *Inst. Adv. Comput.studies*, University of Maryland, College Park, Tech. Rep. UMIACS-TR-96-76 and CSTR-3706, 1996.

[LIN 04] LIN C., PENG H., GRIZZLE J., "Power management strategy for a parallel hybrid electric truck", *IEEE Transactions on Control Systems Technology*, vol. 11, no. 6, pp. 839–849, 2004.

[LIN 11] LIN X., YANG Z., SONG Y., "Intelligent stock trading system based on improved technical analysis and Echo State Network", *Expert Systems with Applications*, vol. 38, no. 9, pp. 11347–11354, September 2011.

[LIU 06] LIU Z., YANG L., MAO Z. *et al.*, "Behavior of PEMFC in starvation", *Journal of Power Sources*, vol. 157, pp. 166–176, 2006.

[LUT 04] LUTSEY N., BRODRICK C.-J., SPERLING D., "Oglesby C., Heavy-duty truck idling characteristics: Results from a nationwide truck survey", *Transportation Research Record: Journal of the Transportation Research Board*, no. 1880, pp. 29–38, 2004.

[LUT 05] LUTSEY N., SPERLING D., "Energy efficiency, fuel economy and policy implications", Transportation Research Record, (1941), 8–17. DOI: 10.3141/1941-02, 2005.

[MAA 02] MAASS W., NATSCHLÄGER T., MARKRAM H., "Real-time computing without stable states: A new frame-work for neural computation based on perturbations", *Neural Compututations*, vol. 14, pp. 2531–2560, 2002.

[MAK 97] MAKRIDAKIS S., HIBON M., "ARMA Models and the Box-Jenkins Methodology", *Journal of Forecasting*, vol. 16, pp. 147–163, 1997.

[MAL 99] MALLAT S., *A Wavelet Tour of Signal Processing*, Elsevier, 1999.

[MAL 08] MALLAT S., *A Wavelet Tour of Signal Processing*, Academic Press, 2008.

[MAL 09] MALLAT S., *A Wavelet Tour of Signal Processing, the Sparse Way*, 3rd edition, Academic Press, 2009.

[MAR 12] MARTINENGHI R., RYBALKO S., JACQUOT M. *et al.*, "Photonic nonlinear transient computing with multiple-delay wavelength dynamics", *Physical Review Letters*, 2012.

[MAS 03] MASTEN DA., BOSCO AD., "System design for vehicle applications: Daimler Chrysler", in VIELTSTICH W. LAMM A., GASTEIGER H. (eds), *Handbook of Fuel Cell Technology – Fundamentals, Technology and Applications*, John Wiley & Sons Inc., New York, vol. 4, pp. 714–724, 2003.

[MAT 13] MATRAJI I., LAGHROUCHE S., JEMEÏ S. *et al.*, "Robust control of the PEM fuel cell air-feed system via sub-optimal second order sliding mode", *Applied Energy*, vol. 104, pp. 945–957, April 2013.

[MCC 81] MCCARTY R.D., HORD J., RODER H.M., "Selected properties of hydrogen (Engineering Design Data)", *Center for Chemical Engineering, National Engineering Laboratory, National Bureau of Standards*, Boulder, Colorado, NBS Report No. 168, February 1981.

[MEN 11] MENCH M.M., KUMBUR E.C., VEZIROGLU T.N., *Polymer Electrolyte Fuel Cell Degradation*, Elsevier Science, 2011.

[MIS 07a] MISITI M., MISITI Y., OPPENHEIM G. *et al.*, *Les ondelettes et leurs applications*, Lavoisier (eds.), Paris, 2007.

[MIS 07b] MISITI M., MISITI Y., OPPENHEIM G. *et al.*, *Wavelets and Their Applications*, ISTE Ltd, 2007.

[MOR 06] MORENO J., ORTUZAR M., DIXON J., "Energy-management systems for a hybrid electric vehicle, using ultracapacitors and neural networks", *IEEE, Trans. on Industrial Electronics*, vol. 53, no. 2, pp. 614–623, April 2006.

[MOR 14] MORANDO S., JEMEÏ S., GOURIVEAU R. *et al.*, "Fuel cells remaining useful lifetime forecasting using echo state network", in *Vehicle Power and Propulsion Conference (VPPC'14)*, pp. IS1–4, October 2014.

[MOR 15a] MORANDO S., Contribution au pronostic de durée de vie d'une pile à combustible à membrane echangeuse de protons, PhD thesis, Université de Franche-Comté, 2015.

[MOR 15b] MORANDO S., JEMEÏ S., HISSEL D. *et al.*, "ANOVA method applied to PEMFC ageing forecasting using an Echo State Network", *Mathematics and Computers in Simulation Journal*, 2015.

[MOR 17] MORANDO S., JEMEÏ S., HISSEL D. *et al.*, "Proton exchange membrane fuel cell ageing forecasting algorithm based on Echo State Network", *International Journal of Hydrogen*, vol. 42, no. 2, pp. 1472–1480, 2017.

[MRÓ 15] MRÓZ W., BUDNER B., TOKARZ W. *et al.*, "Ultra-low-loading pulsed-laser-deposited platinum catalyst films for polymer electrolyte membrane fuel cells", *Journal of Power Sources*, vol. 273, pp. 885–893, doi:10.1016/j.jpowsour.2014.09.173, 1 January 2015.

[MUL 08] MULLER A., SUHNER M.C., IUNG B., "Formalisation of a new prognosis model for supporting proactive maintenance implementation on industrial system", *Reliability Engineering & System Safety*, vol. 93, no. 2, pp. 234–253, 2008.

[MUL 09] MULOT J., Contribution à l'analyse et à la caractérisation d'une alimentation auxiliaire de puissance utilisant une pile à combustible à oxyde solide (APU SOFC) pour les applications transport, PhD thesis, Université de Technologie de Belfort-Montbéliard, 2009.

[MUL 11] MULOT J., JEMEÏ S., PÉRA M.C. *et al.*, "Dynamic and Environmental Performances of Solid Oxyde Fuel Cell Systems fed with natural gas and diesel fuel", *Fundamentals & Developments of Fuel Cells (FDFC)*, S7-O8, France, p. 12, 2011.

[NAU 99] NAUCK D., KRUSE R., "Obtaining interpretable fuzzy classification rules from medical data", *Artificial Intelligence in Medicine*, vol. 16, no. 2, pp. 149–169, June 1999.

[NOT 17] NOTRE PLANÈTE, Hyperlink, available at: https://www.notre-planete.info/ecologie/energie/geothermie.php?54539, 2017.

[OBS 13] OBSERV, "La production d'électricité d'origine renouvelable dans le monde", *EDF*, available at: http://www.energies-renouvelables.org/observ-er/html/inventaire/pdf/15e-inventaire-Chap01-Fr.pdf, 2013.

[ONA 12] ONANENA R., Diagnostic non-intrusif à base de reconnaissance de formes appliqué aux piles à combustible de type PEM, Thesis, Université de Franche-Comté, p. 136, 2012.

[ONG 13] ONGENAE F., VAN LOOY S., VERSTRAETEN D. *et al.*, "Time series classification for the prediction of dialysis in critically ill patients using echo statenetworks", *Engineering Applications of Artificial Intelligence*, vol. 26, pp. 984–996, 2013.

[ORT 07] ORTUZAR M., MORENO J., DIXON J. *et al.*, "Ultracapacitor-based auxiliary energy system for an electric vehicle: Implementation and evaluation", *IEEE, Trans on Industrial Electronics*, vol. 54, no. 4, pp. 2147–2156, August 2007.

[OST 94] OSTWALD W., Die wissenschaftliche Elektrochemie der Gegenwart und die technische der Zukunft, *Z. für Elektrotechnik und Elektrochemie*, vol. 3. pp. 81–84 and 122–125, 1894.

[PAH 14a] PAHON E., JEMEÏ S., HAREL F. *et al.*, "k-Nearest neighbours fault diagnosis of proton exchange membrane fuel cell", *IDHEA Conference*, p. 9, 2014.

[PAH 14b] PAHON E., OUKHELLOU L., HAREL F. *et al.*, "Fault diagnosis and identification of proton exchange membrane fuel cell using electrochemical impedance spectroscopy classification", *ELECTRIMACS Conference*, p. 6, 2014.

[PAH 15a] PAHON E., YOUSFI-STEINER N., JEMEÏ S. *et al.*, "A signal-based method for a proton exchange membrane fuel cell fault diagnosis", *6th International Conference on Fundamentals and Developments of Fuel Cells, FDFC'2015*, Toulouse, France, p. 8, 2015.

[PAH 15b] PAHON E., Contribution au diagnostic de systèmes pile à combustible et accumulateurs electrochimiques, PhD thesis, Université de Franche-Comté, 2015.

[PAH 16] PAHON E., YOUSFI STEINER N., JEMEÏ S. *et al.*, "A signal based method for fast PEMFC diagnosis", *Applied Energy*, vol. 165, pp. 748–758, 2016.

[PAY 03] PAY S., BAGHZOUZ Y., "Effectiveness of battery-supercapacitor combination in electric vehicles", *PowerTech Conference, IEEE*, Bologna, Italy, 23–26 June 2003.

[PEH 03] PEHNT M., "Life cycle analysis of fuel cell system components", in VIELTSTICH W., LAMM A., GASTEIGER H. (eds), *Handbook of Fuel Cell Technology – Fundamentals, Technology and Applications*, John Wiley & Sons Inc., New York, vol. 4, pp. 1293–1317, 2003.

[PEN 14] PENG Z.K., CHU F.L., "Application of the wavelet transform in machine condition monitoring and fault diagnostics: A review with bibliography", *Mechanical Systems and Signal Processing*, vol. 18, pp. 199–221, 2014.

[PÉR 06a] PÉRA M.C., Modélisation de systèmes de conversion d'énergie électrique complexes, HDR, Université de Franche-Comté, 2006.

[PÉR 06b] PÉREZ L., BOSSIO G., MOITRE D. *et al.*, "Optimization of power management in an hybrid electric vehicle using dynamic programming", *Mathematics and Computers in Simulation*, vol. 73, nos 1–4, pp. 244–254, 2006.

[PÉR 13] PÉRA M.C., HISSEL D., GUALOUS H. *et al.*, *Electrochemical Components*, ISTE Ltd, London and John Wiley & Sons, 2013.

[PET 13] PETRONE R., ZHENG Z., HISSEL D. *et al.*, "A review on model-based diagnosis methodologies for PEMFCs", *Int. J. of Hydrogen Energy*, vol. 38, pp. 7077–7091, 2013.

[PRO 14] PROJET DIAPASON2, available at: http://www.agence-nationalerecherche.fr/?Projet=ANR-10-HPAC-0002, 2011–2014.

[PUS 02] PUSCA R., AMIRAT Y., BERTHON A. *et al.*, "Modeling and simulation of a traction control algorithm for an electric vehicle with four separate wheel drives", *Vehicle Technology Conference VTC'02, IEEE*, pp. 1671–1675, 2002.

[RAJ 00] RAJASHERKARA K., "Propulsion system strategies for fuel cell vehicles", *Society of Automotive Engineers*, SAE Paper no. 2000-01-0369, Warrendale, PA, 2000.

[RAU 01] RAUDYS S., *Statistical and Neural Classifiers: An Integrated Approach to Design*, Springer-Verlag, London, 2001.

[RIB 09] RIBOT P., Vers l'intégration diagnostic/pronostic pour la maintenance des systèmes complexes, PhD thesis, Université de Toulouse, 2009.

[ROS 01] ROSSO O.A., BLANCO S., YORDANOVA J. *et al.*, "Wavelet entropy: A new tool for analysis of short duration brain electrical signals", *Journal of Neuroscience Methods*, vol. 105, pp. 65–75, 2001.

[ROS 04] ROSSO O.A., FIGLIOLA A., "Order/disorder in Brain Electrical Activity", *Rev. Mex. Fis*, vol. 50, no. 2, pp. 149–155, 2004.

[RUB 07] RUBIO M.A., URQUIA A., DORMIDO S., "Diagnosis of PEM fuel cells through current interruption", *Journal of Power Sources*, Elsevier, 2007.

[RUB 10] RUBIO M.A., URQUIA A., DORMIDO S., “Diagnosis of performance degradation phenomena in PEM fuel cells”, *International Journal of Hydrogen Energy*, vol. 35, pp. 2586–2590, 2010.

[SAH 09] SAHA B., GOEBEL K., CHRITOPHERSEN J., “Comparision of pronostic algorithms for estimating remaning useful life of batteries”, *Transactions of the Institute of Measurement and Control*, 2009.

[SAI 95] SAITO N., COIFMAN R.R., “Local discriminant bases and their applications”, *Journal of Mathematical Imaging and Vision*, vol. 5, pp. 337–358, 1995.

[SAL 05] SALWANI M., JASMY Y., “Relative wavelet energy as a tool to select suitable wavelet for artificial removal in EEG”, *First International Conference on Computer, Communication and Signal Processing With Special Track on Biomedical Engineering, CCSP*, 2005.

[SAL 07] SALMASI F.R., “Control strategies for hybrid electric vehicles: Evolution, classification, comparison and future trends. IEEE”, *Transaction on Vehicular Technology*, vol. 56, no. 3, pp. 2393–2404, September 2007.

[SAN 14a] SANCHEZ R.E., GOURIVEAU R., JEMEÏ S. *et al.*, “Steiner N. Y. Proton exchange membrane fuel cell degradation prediction based on adaptive neuro-fuzzy inference systems”, *International Journal of Hydrogen Energy*, vol. 39, no. 21, pp. 11128–11144, 2014.

[SAN 14b] SANCHEZ R.S., HAREL F., JEMEÏ S. *et al.*, “Proton exchange membrane fuel cell operation and degradation in short-circuit”, *Fuel Cells Journal*, vol. 14, no. 2, pp. 894–905, DOI: 10.1002/fuce.201300216, 2014.

[SAN 15] SANCHEZ R.E., Contribution au pronostic de durée de vie des systèmes pile à combustible de type PEMFC, Thesis, Franche-Comté et de l’Université de Quebec Trois-Rivières, 2015.

[SCH 08] SCHMITTINGER W., VAHIDI A., “A review of the main parameters influencing longterm performance and durability of PEM fuel cells”, *Journal of Power Sources*, vol. 180, no. 1, pp. 1–14, 2008.

[SCH 12] SCHMIDT K., BOUTALIS Y., “Fuzzy discrete event systems for multi-objective control: Framework and application to mobile robot navigation”, *IEEE Transactions on Fuzzy Systems*, no. 99, 2012.

[SHU 11] SHUMWAY R., STOFFER D., *Time Series Analysis and Its Applications*, Springer New York, 2011.

[SIN 03] SINGHAL S.C., KENDALL K., "High temperature solid oxide fuel cells: fundamentals, design and applications", Elsevier Science Ltd, Oxford, UK, 2003.

[SOL 11] SOLANO J., HISSEL D., PERA M.-C. *et al.*, "Practical control structure and energy management of a testbed hybrid electric vehicle", *IEEE Transactions on Vehicular Technology*, vol. 60, no. 9, pp. 4139–4152, 2011.

[SOL 12a] SOLANO J., JOHN R.I., HISSEL D *et al.*, "A survey based type-2 fuzzy logic system for energy management in hybrid electric vehicles", *Information Sciences*, vol. 190, pp. 192–207, 2012.

[SOL 12b] SOLANO J., Energy management of a hybrid electric vehicle: An approach based on type-2 fuzzy logic, PhD thesis, Franche-Comté University, 2012.

[SOR 09] SORRENTINO M., PIANESE C., "Grey-box modeling of SOFC unit for design, control and diagnostics applications", *European Fuel Cell Forum*, 2009.

[SOR 12] SORENSEN B., *Hydrogen and Fuel Cells, Emerging Technologies and Applications*, 2nd edition, Elsevier Ltd, 2012.

[SOU 13] SOUALHI A., Du diagnostic au pronostic de pannes des entrainements électriques, PhD thesis, Université Claude Bernard Lyon 1, September 2013.

[STE 04] STEIL J.J., "Backpropagation-decorrelation, online recurrent learning with O(N) complexity", *Proceedings of the International Joint Conference on Neural Networks*, vol. 2, pp. 843–848, 2004.

[STO 03] STONE R., "Competing technologies for transportation", in HOOGERS G. (ed.), *Fuel Cell Technology Handbook*, CRC Press, Boca Raton, 2003.

[TAN 08] TANIGUCHI A., AKITA T., YASUDA K. *et al.*, "Analysis of degradation in PEMFC caused by cell reversal during air starvation", *International Journal of Hydrogen Energy*, vol. 33, pp. 2323–2329, 2008.

[TAN 13] TANT S., Contribution au diagnostic d'empilements PEMFC par spectroscopie d'impédance électrochimique et méthodes acoustiques, PhD thesis, Université de Grenoble, July 2013.

[TAW 12] TAWFIK H., HUNG Y., MAHAJAN D., "Chapter 5 – bipolar plate durability and challenges", in MENCH M.M., KUMBUR E.C., VEZIROGLU T.N. (eds), *Polymer Electrolyte Fuel Cell Degradation*, Academic Press, pp. 249–291, 2012.

[TEK 04] TEKIN M., Contribution à l'optimisation énergétique d'un système pile à combustible embarqué, PhD thesis, Université de Franche-Comté, December 2004.

[THO 09] THOUNTHONG P., SETHAKUL P., RAEL S. *et al.*, "Performance evaluation of fuel cell/battery/supercapacitor hybrid power source for vehicle applications", *Industry Applications Society Annual Meeting, 2009. IAS 2009. IEEE*, pp. 1–8, 2009.

[TOY 13] TOYOTA, Toyota global, available at: http://www.toyotaglobal.com/innovation/environmental-technology/technology-file/, 2013.

[UFE 14] UFE, "Le rôle de l'hydrogène dans un mix énergétique décarboné", Observatoire de l'Industrie Électrique, February 2014.

[USD 14] US DEPARTMENT OF ENERGY, Hydrogen. Fuel Cells and Infrastructure Technologies Program, Multi Year Research, Development and Demonstration Plan, Planned program activities for 2005-2015, Technical Plan, available at: http://energy.gov/sites/prod/files/2014/12/f19/fcto_myrdd_fuel_cells.pdf, updated November 2014.

[VER 74] VERNE J., *L'île mystérieuse*, Hetzel, 1874.

[VER 07] VERSTRAETEN D., SCHRAUWEN B., D'HAENE M. *et al.*, "An experimental unification of reservoir computing methods", *Neural Networks*, vol. 20, no. 3, pp. 391–403, 2007.

[VIA 14] VIANNA W.O.L., DE MEDEIROS I.P., AFLALO B.S. *et al.*, "Proton exchange membrane fuel cells (PEMFC) impedance estimation using regression analysis", *IEEE PHM Conference 2014*, pp. 1–8, June 2014.

[VIE 03] VIELSTICH W., LAMM A., GASTEIGER H.A., *Handbook of Fuel Cell – Volume 4: Fundamentals Technology and Applications*, John Wiley & Sons Inc., New York, March 2003.

[VIN 10] VINOT E., TRIGUI R., JEANNERET B., "Optimal management of electric vehicles with a hybrid storage system", *Vehicle Power and Propulsion Conference (VPPC), 2010 IEEE*, pp. 1–6, 2010.

[WAD 10] WADI A., ISMAIL M., KARIM S., "A comparision between Haar wavelet transform and fast Fourier transform in analysing financial time series data", *Research Journal of Applied Sciences*, vol. 5, pp. 352–2360, 2010.

[WAG 04] WAGNER N., GÜLZOW E., "Change of electrochemical impedance spectra (EIS) with time during CO-poisoning of the Pt-anode in membrane fuel cell", *Journal of Power Sources*, vol. 127, pp. 341–347, 2004.

[WAG 09] WAGNER C., HAGRAS H., "Novel methods for the design of general yype-2 fuzzy sets based on device characteristics and linguistic labels surveys", *Proceedings of 2009 IFSA World Congress, Eusflat World Conference*, Lisbon, Portugal, July 2009.

[WAG 10] WAGNER C., HAGRAS H., "Toward general type-2 fuzzy logic systems based on zSlices", *IEEE Transactions on Fuzzy Systems*, vol. 18, no. 4, pp. 637–660, August 2010.

[WAN 11a] WANG H., LI H., YUAN X., *PEM Fuel Cell Durability Handbook*, Taylor & Francis Group, 2011.

[WAN 11b] WANG K., HISSEL D., PÉRA M.C. *et al.*, "A review on solid oxide fuel models", *International Journal of Hydrogen Energy*, vol. 36, pp. 7212–7228, 2011.

[WAN 12] WANG K., Ex-situ and in-situ diagnostic algorithms and methods for solid oxide fuel cell systems, PhD thesis, University of Franche-Comte, p. 168, 2012.

[WAN 13a] WANG H., LISERRE M., BLAABJERG F., "Toward reliable power electronics: Challenges, design tools, and opportunities", *IEEE Industrial Electronics Magazine*, vol. 7, no. 2, pp. 17–26, June 2013.

[WAN 13b] WANG H., ZHOU D., BLAABJERG F., "A reliability-oriented design method for power electronic converters", *2013 Twenty-Eighth Annual IEEE Applied Power Electronics Conference and Exposition (APEC)*, pp. 2921–2928, March 2013.

[WAN 14] WANG H., LISERRE M., BLAABJERG F. *et al.*, "Transitioning to physics-of-failure as a reliability driver in power electronics", *IEEE Journal of Emerging and Selected Topics in Power Electronics*, vol. 2, no. 1, pp. 97–114, March 2014.

[WAS 10] WASTERLAIN S., Approches expérimentales et analyse probabiliste pour le diagnostic de pile à combustible de type PEM, Thesis, Université de Franche-Comté, 2010.

[WIA 00] WIARTALLA A., PISCHINGER S., BORNSCHEUER W. *et al.*, "Compressor expander units for fuel cell systems", *Society Automotive Engineers Inc.*, 2000.

[WIL 08] WILTCHKO A., GAGE G., BERKE J., "Wavelet transform before spike detection preserves waveform shape and enhances singleunit discrimination", *Journal of Neurosciences Methods*, vol. 173, pp. 34–40, 2008.

[WU 05] WU J.-D., CHEN J.-C., "Continuous wavelet transform technique for fault signal diagnosis of internal combustion engines", *NDT&E International Pages*, pp. 1–8, 2005.

[WU 08a] WU J., YUAN X.Z., MARTIN J.J. *et al.*, "A review of pem fuel cell durability: Degradation mechanisms and mitigation strategies", *Journal of Power Sources*, vol. 184, no. 1, pp. 104–119, 2008.

[WU 08b] WU J., YUAN X.Z., WANG H. *et al.*, "Diagnostic tools in PEM fuel cell research: Part I Electrochemical techniques", *International Journal of Hydrogen Energy*, vol. 33, pp. 1735–1746, 2008.

[WU 08c] WU X.-J., ZHU X.-J., CAO G.-Y. *et al.*, "Nonlinear modeling of a SOFC stack based on ANFIS identification", *Simulation Modelling Practice and Theory*, vol. 16, no. 4, pp. 399–409, 2008.

[WUR 97] WURSTER R., PEM Fuel Cells in stationary and mobile applications, Biel conference, available at: http://www.hyweb.de/knowledge, 1997.

[YAN 07] YAN R., Base wavelet selection criterion for non-stationnary vibration analysis in bearning health diagnosis, Thesis, University of Massachussets Amherst, 2007.

[YAO 13] YAO B., CHEN H., HE X.-Q. *et al.*, "Reliability and failure analysis of DC/DC converter and case studies", *2013 International Conference on Quality, Reliability, Risk, Maintenance, and Safety Engineering (QR2MSE) IEEE*, pp. 1133–1135, July 2013.

[YOU 09a] YOUSFI-STEINER N., MOÇOTÉGUY P., CANDUSSO D. *et al.*, "A review on polymer electrolyte membrane fuel cell catalyst degradation and starvation issues: Causes, consequences and diagnostic for mitigation", *Journal Power Sources*, vol. 194, no. 1, pp. 130–145, October 2009.

[YOU 09b] YOUSFI-STEINER N., HISSEL D., CANDUSSO D. *et al.*, Détection de défaut dans un dispositif électrochimique; brevet d'invention; demande prioritaire: FR 09 54357; Extension de la protection N: PCT/FR2010/051295, patent N° US20120116722, 25 June 2009.

[YOU 09c] YOUSFI-STEINER N., Diagnostic non intrusif de groupes électrogènes à piles à combustible de type PEMFC, PhD thesis, Université de Franche-Comté, 2009.

[YOU 11a] YOUSFI-STEINER N., HISSEL D., MOÇOTEGUY P. *et al.*, "Non intrusive diagnosis of polymer electrolyte fuel cells by wavelet packet transform", *International Journal of Hydrogen Energy*, vol. 36, pp. 740–746, 2011.

[YOU 11b] YOUSFI-STEINER N., HISSEL D., MOÇOTÉGUY P. *et al.*, "Diagnosis of polymer electrolyte fuel cells failure modes (flooding & drying out) by neural networks modeling", *Int. J. of Hydrogen Energy*, vol. 36, no. 4, pp. 3067–3075, 2011.

[YUA 07] YUAN X., WANG H., SUN J.C. *et al.*, "AC impedance technique in PEM fuel cell diagnosis-A review", *International Journal of Hydrogen Energy*, vol. 32, pp. 4365–4380, 2007.

[YUA 12] YUAN X.Z., ZHANG S., BAN S. *et al.*, "Degradation of a PEM fuel cell stack with Nafion® membranes of different thicknesses. Part II: Ex situ diagnosis", *Journal of Power Sources*, vol. 205, pp. 324–334, May 2012.

[ZAD 65] ZADEH L., "Fuzzy sets", *Information and Control*, vol. 8, pp. 338–353, 1965.

[ZHA 12] ZHANG X., PISU P., "An unscented kalman filter based approach for the healthmonitoring and prognostics of a polymer electrolyte membrane fuel cell", *Proceedings of the Annual Conference of the Prognostics and Health Management Society*, 2012.

[ZHE 13a] ZHENG Z., PETRONE R., PÉRA M.C. *et al.*, "A review on non-model based diagnosis methodologies for PEM fuel cell stacks and systems", *International Journal of Hydrogen Energy*, pp. 1–13, 2013.

[ZHE 13b] ZHENG Z., PETRONE R., PÉRA M.C. *et al.*, *39th Annual Conference of the IEEE Industrial Electronics Society (IECON)*, pp. 1595–1600, 2013.

[ZHE 14] ZHENG Z., On-line faul diagnosis of PEMFC stacks via on non-model based methods and EIS measurements, Thesis, Université de Franche-Comté, September 2014.

Index

A, C, D

E, F, G

M, N, O, P

S, T, U, W

Other titles from

iSTE

in

Energy

2018

BENALLOU Abdelhanine
Energy and Mass Transfers: Balance Sheet Approach and Basic Concepts
(Energy Engineering Set – Volume 1)
Energy Transfers by Conduction
(Energy Engineering Set – Volume 2)

RUFFINE Livio, BROSETA Daniel, DESMEDT Arnaud
Gas Hydrates 2: Geoscience Issues and Potential Industrial Applications

VIALLET Virginie, FLEUTOT Benoit
Inorganic Massive Batteries
(Energy Storage – Batteries, Supercapacitors Set – Volume 4)

2017

BROSETA Daniel, RUFFINE Livio, DESMEDT Arnaud
Gas Hydrates 1: Fundamentals, Characterization and Modeling

LA SCALA Massimo
From Smart Grids to Smart Cities: New Challenges in Optimizing Energy Grids

MOLINA Géraldine, MUSY Marjorie, LEFRANC Margot
Building Professionals Facing the Energy Efficiency Challenge

SIMON Patrice, BROUSSE Thierry, FAVIER Frédéric
Supercapacitors Based on Carbon or Pseudocapacitive Materials
(Energy Storage – Batteries, Supercapacitors Set – Volume 3)

2016

ALLARD Bruno
Power Systems-on-Chip: Practical Aspects of Design

ANDRE Michel, SAMARAS Zissis
Energy and Environment

DUFOUR Anthony
Thermochemical Conversion of Biomass for the Production of Energy and Chemicals

2015

CROGUENNEC Laurence, MONCONDUIT Laure, DEDRYVÈRE Rémi
Electrodes for Li-ion Batteries
(Energy Storage – Batteries, Supercapacitors Set – Volume 2)

LEPRINCE-WANG Yamin
Piezoelectric ZnO Nanostructure for Energy Harvesting
(Nanotechnologies for Energy Recovery Set – Volume 1)

ROBYNS Benoît, FRANÇOIS Bruno, DELILLE Gauthier, SAUDEMONT Christophe
Energy Storage in Electric Power Grids

ROSSI Carole
Al-based Energetic Nanomaterials
(Nanotechnologies for Energy Recovery Set – Volume 2)

TARASCON Jean-Marie, SIMON Patrice
Electrochemical Energy Storage
(Energy Storage – Batteries, Supercapacitors Set – Volume 1)

2013

LALOUI Lyesse and DI DONNA Alice
Energy Geostructures: Innovation in Underground Engineering

2012

BECKERS Benoit
Solar Energy at Urban Scale

ROBYNS Benoît, DAVIGNY Arnaud, FRANÇOIS Bruno, HENNETON Antoine, SPROOTEN Jonathan
Electricity Production from Renewable Energies

2011

GAO Fei, BLUNIER Benjamin, MIRAOUI Abdellatif
Proton Exchange Membrane Fuel Cell Modeling

MULTON Bernard
Marine Renewable Energy Handbook

2010

BRUNET Yves
Energy Storage

2009

SABONNADIÈRE Jean-Claude
Low Emission Power Generation Technologies and Energy Management

SABONNADIÈRE Jean-Claude
Renewable Energy Technologies

Printed and bound by CPI Group (UK) Ltd, Croydon, CR0 4YY

16/09/2026

14963074-0001